Innovation and Entrepreneurship in Western Canada

Innovation and Entrepreneurship in Western Canada: From Family Businesses to Multinationals

edited by

James J. Chrisman
University of Calgary

J. Adam D. Holbrook
Simon Fraser University

Jess H. Chua
University of Calgary

UNIVERSITY OF CALGARY PRESS

University of Calgary Press
2500 University Drive N.W.
Calgary, Alberta T2N 1N4
www.uofcpress.com

National Library of Canada Cataloguing in Publication Data

Main entry under title:

Innovation and entrepreneurship in Western Canada

Papers from the conference, InnoComm, held at the University of Calgary, Apr. 27-28, 2000.
Includes bibliographical references and index.
ISBN 1-55238-075-0

1. Research, Industrial—Canada, Western—Congresses. 2. Technological innovations—Economic aspects—Canada, Western—Congresses. 3. Entrepreneurship—Canada, Western—Congresses. 4. Strategic alliances (Business)—Canada, Western—Congresses. 5. Canada, Western—Economic conditions—Congresses. I. Chrisman, James J. II. Holbrook, J. Adam, 1946- III. Chua, Jess H.
HC120.T4I585 2002 338'.064'09712 C2002-910247-2

Canada We acknowledge the financial support of the Government of Canada through the Book Publishing Industry Development Program (BPIDIP) for our publishing activities.

The Canada Council for the Arts
Le Conseil des Arts du Canada

Printed and bound in Canada by AGMV Marquis.
∞ This book is printed on acid-free paper.

Cover design by Sona Khosla.
Typesetting and page design by Cliff Kadatz.

Contents

Preface

This book presents a series of articles on innovation, entrepreneurship, and family business from a western Canadian perspective. It deals with the characteristics of the innovation and entrepreneurship processes, and how they can be facilitated. Furthermore, it introduces into the equation a type of organization, family business, that has received scant attention in the literature despite the prominence and importance of this organizational form. A unique aspect is the book's western Canadian focus. Innovation, entrepreneurship, and family business are first and foremost regional phenomena that are crucial to economic development. Many of the articles are based on research or circumstances that are distinctly western Canadian in their flavor and use western Canadian examples and data for illustration and evidence. In addition, all of the articles are authored or co-authored by scholars and/or practitioners who live, work, and are affected by the western Canadian approach to economic systems.

This book deals with a number of disciplines. Its contents fall primarily within the field of entrepreneurship, since that field encompasses all of the major subject areas covered. However, the book should appeal to any reader concerned with innovation, economic development, or family business. Thus, scholars in fields as diverse as sociology, economics, strategic management, marketing, public administration, finance, communications, medicine, technology management, etc., should find the contents valuable.

The plan for the book originated with the preparations for the third conference of InnoCom, the western Canadian sub-network of the Innovation Systems Research Network (ISRN) in Canada. The ISRN is funded by the National Research Council (NRC), the Social Sciences and Humanities Research Council (SSHRC), and the Natural Sciences and Engineering Research Council (NSERC) expressly for the purposes of developing networks of scholars interested in innovation, fostering exchanges of ideas, and encouraging collaborative efforts. The papers included in this book were presented at the third InnoCom conference at the Faculty of Management of the University of Calgary on April 27 and 28, 2000.

This conference differed from the first two in that it deliberately sought to broaden the scope of inquiry into innovation and its antecedents. Thus, while the first two conferences provided valuable insights into the process of innovation, we felt something was missing. What was missing was consideration of the people who innovate and those new, entrepreneurial firms that contribute so much to the innovation system in any region or nation. It also occurred to us that many entrepreneurial firms begin as, or develop into, family businesses. We further noted that although they are a global phenomenon, family businesses represent a particularly important part of the economic fabric of western Canada. As a consequence, we decided that the theme "Innovation, Entrepreneurship, Economic Development, and Family Business: A Western Canadian Perspective" would be appropriate in achieving our goals for breadth while maintaining the integrity of the purpose of the InnoCom organization. Scholars from outside InnoCom were sought from all of the leading universities in western Canada. All were informed of the conference's theme, its expanded scope, and its mandate to integrate the topic areas of innovation, entrepreneurship, economic development, and family business. Our intention was, to the extent possible, to have a group of papers each of which dealt with at least two of the conference topics. By and large, we feel that this goal was achieved, an impressive outcome considering that the mix of papers had to be partially a function of the research interests and programs of the contributors.

By bringing together scholars with different perspectives on these topics we hope to emphasize the relationships among them, to alert scholars to different aspects of the economic development problem, and to facilitate the creation of networks of scholars and research programs. In other words, this book provides a glimpse into the future direction of research among the faculty at major research institutions of western Canada including the University of Alberta, the University of Calgary, the University of Lethbridge, the University of Manitoba, the University of Saskatchewan, Simon Fraser University, and the University of Victoria.

The editors hope this book demonstrates the benefits of interdisciplinary studies. Of course we expected that the conference would be a success, but never in our wildest dreams could we have conceived that bringing together scholars from such a wide array of disciplines and with such diverse interests would be so productive. On the surface it might appear that even the combination of scholars from business schools specializing in fields such as finance, strategic management,

management information systems, and organization behavior would be too diverse a mix to yield benefits, despite their shared interests in innovation, entrepreneurship and family business. To then add scholars from interdisciplinary research programs rooted in applied political theory and geography as well as those with backgrounds in communications, economics, and chemistry appears to create a strange brew indeed. But in fact, the mix reinforced a basic premise of the meeting: that the study of innovation, entrepreneurship, and family business in western Canada (or anywhere else for that matter) will be more productive when people from the widest possible set of orientations are brought together to exchange ideas. Although each chapter is a standalone contribution to the literature, we hope that the breadth and depth of the topic areas demonstrate both the opportunities and challenges that lie ahead as we attempt to gain a better understanding of how innovation, entrepreneurship, and family business management influence economic development.

Acknowledgments

A project of this magnitude could have never been completed without the help of a large number of dedicated individuals. The editors and authors would like acknowledge their efforts.

First, we thank our sponsors – the University of Calgary's Centre for Family Business Management and Entrepreneurship, InnoCom, and the Government of Alberta – for their generous support of the project. The sponsorship of the Centre for Family Business Management and Entrepreneurship in the Faculty of Management at the University of Calgary has been made possible by an endowment from an anonymous benefactor. InnoCom is an interdisciplinary network of researchers in western Canada, joined together by a common interest in studying the dynamics of innovation. In addition to the direct SSHRC grant to the network (#822-98-0005), each researcher had his or her own sources of support. These sources include, but are not limited to: Statistics Canada, Indian Affairs Canada, Industry Canada, Western Economic Diversification Canada, the Science Council of B.C., the B.C. Advanced Systems Institute, the Alberta Department of Innovation and Science, the National Research Council of Canada, the Social Sciences and Humanities Research Council of Canada, and the Small Business Development Center Program in the United States. A generous donation from the Government of Alberta was made possible through the support and friendship of the Honourable Murray Smith, MLA and Minister of Gaming.

Second, thanks are due to Judy McCaffrey, Lori Gibson, and Karen Chrisman of the Faculty of Management of the University of Calgary for their help in organizing the meeting.

Third, a special acknowledgement must be made of Martha Langford, of the Science and Technology in Society program of the Faculty of Communications and Culture at the University of Calgary, for her assistance in managing the internal peer review process of the individual manuscripts. We also wish to thank the external reviewers of the entire manuscript as well as the following individuals who, prior to its submission, each reviewed one of the papers that make up the chapters in the book:

Mike Chiasson, University of Calgary
Peter Josty, Josty and Associates
Cooper H. Langford, University of Calgary
Martha W. Langford, University of Calgary
Reginald Litz, University of Manitoba
W. Ed McMullan, University of Calgary
James O'Grady, University of Calgary
Tim Padmore, University of British Columbia
Yun Park, University of Lethbridge
Peter W.B. Phillips, University of Saskatchewan
Peter Robinson, University of Calgary
Richard Smith, Simon Fraser University
Myron Weber, University of Calgary

Finally, we wish to thank the authors of the papers for their time and efforts. There are excellent scholars in western Canada, and we believe the contributors to this book to be among the best one will find anywhere.

Introduction

Chapter 1

Introduction to Innovation and Entrepreneurship in Western Canada: From Family Businesses to Multinationals

James J. Chrisman, Jess H. Chua, and J. Adam D. Holbrook

Economic development is about regional or national economic changes that lead not only to growth but to improvements in the capacities and capabilities of an economic system (Reese and Fasenfast 1997). A central thesis of this book, as reflected in this introduction and the following chapters, is that innovation and entrepreneurship are the driving forces of economic development and that they occur within new and old, large and small, public and private organizations. Although the book is concerned primarily with innovation and entrepreneurship in western Canada, the issues, insights, and opinions are pertinent to other regions and national economies.

The book seeks to enrich our understanding about how innovation and entrepreneurship may be increased or made more effective and efficient. It stresses the importance of knowledge creation, people, trust, a network of relationships, and critical mass as the ingredients that determine the level of innovation and entrepreneurship, whether it be in family businesses or multinational corporations. Each chapter deals with one or more of these topic areas.

As noted in the Preface, because the book is designed to cover a wide range of subjects from a wide range of perspectives using a wide range of theoretical and methodological approaches, any reader concerned with innovation, entrepreneurship, economic development, or family business should find something of value within its pages. By bringing together scholars with different perspectives on these topics, we hope to emphasize the relationships among them, alert scholars to different aspects of the economic development problem, demonstrate the benefits of cross-disciplinary research, and facilitate the creation of networks of scholars and research programs.

In this introduction we highlight the contributions made by each chapter and discuss how the chapters are interrelated within the overall structure of the book. To achieve these aims this introduction chapter is organized as follows. First, we summarize the situation in western Canada to provide perspective. Second, we give a brief introduction to the topics of innovation, entrepreneurship, and family business. Third, we provide brief synopses of the four sections and the fifteen chapters that comprise the remainder of the book. Finally, we identify and discuss the underlying themes that link the contributions of the fifteen chapters.

The Western Canadian Context

In western Canada, innovation has become critically important as the economies of Alberta, British Columbia, Manitoba, and Saskatchewan try to reduce their dependence on resource-based industries such as agriculture, forestry, mining, and petroleum. This will not be easy, because in the early days of this country, western Canada's historic role in the national economy was to contribute resource and agricultural outputs to Ontario and Quebec and to consume manufactured products from their industries, many of which were protected from international competition by high tariffs. The removal of trade barriers has forced eastern Canada's manufacturing industries to become globally competitive and less reliant on tariffs for survival. Industries that have not become so have disappeared or are disappearing. This has not happened for the western provinces' resource-based businesses, most of which are now in vicious competition with producers around the world.

Although in the first years of the new millennium the economies of western Canada have been relatively healthy and are becoming more diversified, over the past few decades western Canadian resource-based products have lost ground to external competition. Wheat production in the Prairies has suffered from Canada's inability to compete with subsidies given to U.S. and EU farmers. Forest products in British Columbia have to compete with softwood from third world countries while transportation over longer and longer distances between forests and mills continue to add to costs. Intensive livestock industries in Alberta, Saskatchewan and Manitoba are running into community resistance for environmental reasons. Prices for gold and base metals have remained low, making low-concentration ore bodies uneconomical. Only through innovation can the region leave its past behind, reduce its reliance on commodity industries, and escape the slowly deteriorating competitive situation into which its economy is sliding. Thus, to western Canada, innovation is not only an opportunity but also a critical necessity.

Entrepreneurs and entrepreneurial firms are central to the process of innovation and key to the future economic development of western Canada. New firms that add value to raw materials rather than just extract them will be essential. Fortunately, entrepreneurship is alive and well in western Canada. In fact, western Canada's resource industries have fostered a great deal of entrepreneurship in new and traditional fields.[1] The region is also home to several important organizations that facilitate the linked processes of innovation and entrepreneurship.[2] For example, the Canadian Venture Exchange, an exceedingly innovative stock exchange designed to provide capital for early stage companies, has been very successful in helping to launch many companies set up to exploit innovative technologies. The region is also home to the two most successful technology transfer organizations in Canada, one at the University of Calgary and the other at the University of Alberta. And as the articles in Chapters 9 and 11 (Section III) suggest, western Canada is also the home to many of the national innovations occurring in entrepreneurship education. While all such activity provides a sound basis for the future, if the region is to flourish the entrepreneurial spirit of the region must be maintained, support programs further strengthened, and venture capital funding enhanced.

Western Canada has a higher proportion of small and medium-sized enterprises than the manufacturing heartland of Ontario and Quebec. Although Calgary is now ranked second in the nation in terms of corporate headquarters, most of the largest Canadian corporations remain headquartered in the east, particularly in Toronto. It appears that medium-sized enterprises in western Canada are far more likely to be family-owned and operated. For example, Slator (see Chapter 12) points out that, among his venture capital company's investments, only those in Alberta involve family firms; not a single one of their investments in eastern Canada or the U.S. is in family-controlled firms. The behavior of these enterprises, and the behavior of family-owned enterprises, is a matter of special interest in western Canada.

Another factor that appears to set western Canada apart is the importance of defining a role for its aboriginal people. In 1999, the Canada West Foundation conducted a survey to determine the most important questions on the western Canadian economic research agenda. Trade and economic growth received the highest rating. However, the socio-

[1] See, for example, Phillips' case study of biotechnology in Saskatoon's canola industry (Chapter 2).

[2] See, for example, the articles by Josty (Chapter 4) and Langford, Tyrie, and Peace (Chapter 8).

economic prospects for aboriginal peoples received a higher rating by western Canadian policy makers than by those in other parts of Canada. There are many pending issues involving aboriginal peoples, such as land claims, transportation, support for remote communities, and the trade-off between resource exploitation and environmental protection. In Manitoba and Saskatchewan, members of the First Nations are soon to become the majority of the population. As Mitchell and Morse illustrate (Chapter 6), innovation and entrepreneurship may offer a solution to the economic development needs of these people, particularly in western Canada.

In summary, the western Canadian economies have historically been dominated by resource industries. If the region is to break out of its dependency on such industries, in which it is losing its global competitiveness, on the one hand, and capture a greater share of the value-added potential within these industries on the other, innovation will be critically important. Because of the structural bases of western Canadian industries, innovation is likely to come primarily from entrepreneurial firms, and many of these firms will be family-owned. Fortunately, the region is well situated to support such ventures. Finally, an important additional concern for economic development in Western Canada is the involvement of the aboriginal communities. The largely underutilized human resources available in these communities need to be mobilized. Again, innovation and entrepreneurship seem to be the keys to achieving this objective.

Innovation, Entrepreneurship, and Family Business

Innovation

To paraphrase Rogers (1962), an innovation is the development of a new productive use for ideas, discoveries, or resources. Success in innovation rests more with the commercial success of an idea than on its technical or social merits (Burgelman and Sayles 1986). As such, an innovation's contribution to economic development tends to be of three types. First, innovations may lead to the development of new products or services, such as computers, fax machines, or overnight courier services. Second, innovations may lead to improvements in the function or quality of existing products or services such as the application of integrated circuits and digital memory to photocopying, or discount brokerage services. Third, innovations can lead to the application of new, more efficient processes for the production or distribution of products or services, as in the float technology used to produce glass, continuous casting of steel (Utterback 1994), or internet auctions. In innovations can emanate from any link within the value material extractor, intermediate supplier, manufacturer, distributor, retailer, support service provider, or end user (von Hippel 1988).

Although the significance of the productive use is mainly measured in economic terms, innovations have other important impacts on society. Innovations have improved our health and longevity (pacemakers, antibiotics), increased our access to information (the internet), and made communication and travel easier (cellular phones and fuel-efficient jet aircraft). Innovations have also created new leisure activities (internet chat and snowboarding) and increased the amount of time available to engage in such activities (microwave ovens and washing machines). In fact, the economic benefits of most innovations would generally not occur without corresponding social impacts. Thus, while innovations are supposed to create new jobs and lead to higher profits, neither would occur without direct or indirect social benefits that make the resulting economic activities worthwhile to people. In fact, profits measure the additional value society places on those products or services compared to the cost of the resources (or alternative uses for those resources) used to make them ready for consumption. New jobs are created because people are willing to purchase products and services at a sufficient price in a sufficient quantity to justify increases in the production of those products and services. Put simply, innovations must have demonstrable social value in order to have lasting economic value.

Entrepreneurship

Sharma and Chrisman (1999, 17) define entrepreneurship as "acts of organizational creation, renewal, or innovation that occur within or outside of existing organizations." The contributions of entrepreneurship to economic development are well known. Small and new firms account for a disproportionate amount of the jobs created in the economy (Birch 1987) and a disproportionate number of the innovations in many industrial categories (Acs and Audretsch 1990). As Soukup (1979) notes, firms that successfully introduce major new technologies almost always come from outside traditional industries. It is these entrepreneurial firms that capture our major interest.

According to the above definition, innovation is an entrepreneurial act in that it involves the commercialization of new combinations of resources with economic value (Schumpeter 1934). Such acts may or may not also lead to the creation of new organizations. However, entrepreneurship is not necessarily innovative; organizations may be created or re-created without significant innovative activity.[3] The new convenience store down the block would not normally be considered an innovative organization although it is bound to possess its own unique features. While we are obviously most interested in the creation of innovative organizations, we would be remiss if we did not point out that imitative organizations also play an important role in society. Intel cannot produce all of the world's chips, Microsoft cannot (and will not be allowed to) produce all of the world's software, Amazon cannot sell all of the world's books, and McCain's cannot make all of the world's frozen French fries. And no one can deny the economic or social value of companies that manufacture generic drugs.

It is important to note that while the physical manifestation of entrepreneurship is usually a new organization (and/or new products or processes), the entrepreneur is the embodiment of entrepreneurship (Bygrave and Hofer 1991). It is the confluence of behavior, strategic decisions, and skills an entrepreneur uses to start and operate a venture that determine what the venture will be and whether it will be successful (Chrisman, Bauerschmidt, and Hofer 1998; Gartner 1988; Herron and Robinson 1993). Therefore, as we discuss entrepreneurship and innovation it is important not to lose sight of the human actors whose choices drive the processes and results.

Family business

There is no consensus on the definition of a family business. In the past, researchers emphasized the components of family involvement: ownership, management, and succession within the family. Recently, the emphasis has shifted toward definitions that capture the "essence" of a family business. The components are seen as the tools that, when applied appropriately, make the essence possible. Chua, Chrisman, and Sharma (1999, 25) suggest that family businesses are "governed and/or managed with the intention to shape and pursue the vision of the business held by a dominant coalition controlled by members of the same family or a small number of families in a manner that is potentially sustainable across generations of the family or families." In other words, family businesses are those organizations wherein members of the same family control the strategic direction of the business.

---

[3] Holbrook and Hughes' study (Chapter 7) shows that innovative organizations display some markedly different organizational attributes than less- or non-innovative organizations.

Many new firms, whether innovative or not, eventually become family firms. This is why many believe that most businesses are family businesses. Although precise national statistics cannot be obtained, estimates of their predominance range from 65% to 90% of all businesses (Beckhard and Dyer, 1983). Shanker and Astrachan (1996) estimate that there may be as many as 20.3 million family businesses in the U.S., accounting for up to 49% of GDP and 59% of total employment. Similar figures have been reported for Canada, where family firms are found to provide jobs for over 6 million Canadians (Deloitte and Touche 1999). Therefore, it is undeniable that family businesses represent a significant portion of the world economy.

The value that family businesses add to the world economy is only beginning to be understood. In general, economic transactions are effected using markets, hierarchies, and/or relational contracts (Williamson 1985). Family involvement has significant implications for both hierarchies and relational contracts. Family hierarchies naturally play a role in the family firm. Relational contracts, whether explicit or implicit, depend critically on trust, of which families, hopefully, have more.[4] Shleifer and Vishny (1997) argue that when the legal and political institutions needed to control the agency costs involved in balancing the interests of ownership and management are missing, the family firm is an efficient organizational form.[5]

People do not naturally associate family businesses with entrepreneurial or innovative behaviors. Yet if these organizations are to realize the family's vision for the business, they must be able to compete, and there is no more sustainable competitive advantage than continual innovation or entrepreneurship. For example, Tan and Fock (2001) argue that only those Singaporean Chinese firms with entrepreneurial successors were able to make the transition into global companies.

There are, in fact, many examples of family firms that have succeeded through innovation and entrepreneurship. Wal-Mart was the first to use information technology to minimize its inventory level and thus create a cost advantage compared to its competitors. Canadian Tire attributes the success of its network of dealerships to its use of information technology to design a different product mix for each of its far-flung outlets. Johnson's Wax recently introduced new mops and cleaning

[4] Steier's paper (Chapter 16) provides an interesting overview of trust and its relationship to organizational evolution in the family firm.

[5] Agency theory has also been found to be useful in explaining the behaviors of family firms. For example, Park's article (Chapter 15) shows that agency issues helps to explain the compensation practices of family firms in Canada.

fluids to take advantage of the surge in popularity of hardwood flooring. Cargill's meat processing plant in High River, Alberta is able to track, in real time, the cost of each box of meat going out the plant gate.

By including family business topics in this book about innovation and entrepreneurship, we extend our understanding of economic development beyond the founder generation of innovative and entrepreneurial firms. The potential contributions that these firms can make past the first generation depend critically on their ability to deal with the special problems associated with the interaction between family and business concerns. In this vein, Litz and Kelysen's article (Chapter 13) provides perhaps the first detailed discussion of the factors influencing transgenerational innovation in family firms.

Traditionally, scholars have assumed that these concerns most strongly manifest themselves in issues such as leadership succession, strategic planning, professionalization, conflict resolution, and governance through the board of directors (Sharma, Chrisman, and Chua, 1997). In fact, the family business literature has been dominated by studies of succession. Chua, Chrisman, and Sharma's article (Chapter 14) shows that this emphasis in not without merit, but adds relationships with non-family managers as another critical concern.

An Overview of the Book

Section I: Innovation systems

Given the economic situation facing western Canada, it is appropriate that our first set of chapters deals with systems of innovation. These chapters provide both a conceptual and empirical understanding of innovation systems as they apply to western Canada, using both case study and interview methodologies. They clearly suggest that the path to progress depends on a region's ability to become integrated into the larger national and international systems. Smith (Chapter 3) provides additional value by describing how to map the innovation system in a region. Even though the western Canadian system is unique in many respects, we believe that these chapters will provide insight for researchers regardless of geographic focus, since there are many commonalities between the features of the western Canadian systems and other regions.

Chapter 2: Regional systems of innovation as modern R&D entrepots: The case of the Saskatoon biotechnology cluster

In this chapter, Peter Phillips synthesizes and expands on alternative theoretical approaches that view innovation from either a firm-level or national systems perspective. He argues that modern innovation develops with and among regional firms and agencies that are integrated into a global production system. In his case study, Phillips notes that Saskatoon has developed a highly advanced canola industry based on a system that reflects many of the characteristics of the classic trade entrepot, where inputs are brought in tax-free, local value is added, and outputs are exported for further value-added processing and sale to final customers. In the canola industry, basic research emanates primarily from the U.S., with the local industry in Saskatoon being dominant in variety breeding, field trials, and primary production. However, most of the activities that occur downstream are conducted elsewhere.

The canola case illustrates how a local presence can be maintained and integrated into a global system of innovation and production. Phillips' theory is intriguing enough to warrant further development and testing to determine the extent to which it holds in other industries and for other technologies. As he suggests, to be able to make appropriate public policy recommendations with regard to knowledge transfers and trade, additional case studies would also be useful. We need to gain a better understanding of how and how well modern entrepots work, and where value is created and captured, if international systems of innovation are to flourish.

Chapter 3: Techmaps: A tool for understanding social capital for technological innovation at a regional level

One of the most difficult problems facing researchers in entrepreneurship and innovation studies is identifying the relevant population of firms and tracking their contributions. Even more difficult is identifying and understanding the contributions of firms that no longer exist. Solving this problem is the first step toward understanding how innovations diffuse throughout the economy. Without such understanding, developing a generalizable body of knowledge about innovation will continue to be a daunting task.

Chapter 3, by Richard Smith, presents one of the most ambitious attempts to date to deal with this issue. Smith discusses the labor-intensive process by which he mapped the firms in the technology sector of a region. His chapter demonstrates that it is possible to identify the population, trace the genealogy of individual companies, and attribute the contributions made by the firms.

One can learn much from the successes and difficulties he describes in the chapter. Furthermore, existing and future techmaps offer a rich source of data for developing and testing theories on the timing, geographic scope, and the nature of innovations emanating from a region, as well as insights into the question of where entrepreneurs come from and how new ventures are formed. We see the techmap project as a first step toward a taxonomic research program akin to the identification of species in biology. Such work will make more effective functional studies of innovation and entrepreneurship possible.

Chapter 4: The Alberta innovation system: The main players and suggested research priorities

Peter Josty's chapter provides a highly informative description of the complex web of players who contribute to innovation in a region. It discusses governmental bodies that set policy, regulations, and taxes, generators of knowledge, innovators, sources of capital, service providers, and educational institutions. His analysis provides perspective as to how well Alberta is doing in meeting its goals with resepect to innovation as well as the gaps that need to be closed in its innovation system.

For example, Josty suggests that changes to income tax provisions on the sale of stock options, greater willingness to take risks, and reduced regulatory barriers would increase the level of entrepreneurship and innovation in the province. He ends by proposing 10 research priorities for future studies of innovation in Alberta, including gaining a better understanding of business practices that encourage innovation, innovation in commodity industries, industry cluster formation, diffusion of innovation, and how research funds create economic value.

Chapter 5: Cooperative mercantilism: Public policy dilemmas managing the generation and application of ideas toward economic development

In the final chapter of this section, Michael Chiasson provides a broad context for discussing innovation systems. He seeks to address two important questions: (1) What is the relationship between science and innovation both globally and regionally? and (2) What role can public policy play in supporting the development and use of regional and global knowledge in local economic development? Drawing on the ideas of authors such as Phillips (Chapter 2), Chiasson argues that effective systems of innovation in a nation or region must achieve a balance between a free and borderless exchange of ideas and the protection of the competitive advantages accruing from the ideas generated within a region. Such a balance, which he dubs "cooperative mercantilism", fosters an environment for both basic and applied research, allows knowledge to be exchanged and used in a way that develops local advantages through strategic linkages, and creates strong social networks with regional and global communities.

Universities, Chiasson believes, will play an increasingly important role in achieving a balance. Interestingly, he suggests that because of the challenges inherent in increased knowledge specialization, and due to the regional roots of both types of institutions, family businesses, in partnership with universities, will also be important players. We would add that with the right leadership in place, the overlaps between ownership and management typically found in family firms could give them further advantages in making timely responses to technological change and in utilizing the products of science to innovate.

Section II: Entrepreneurship and Innovation

The second section of this book provides three chapters that deal with the intersection between innovation and entrepreneurship. Whereas the first section deals with innovation at the system level, this section examines it at the level of individual social, business, and support organizations. The focus in both cases is on increasing the level, as well as improving the effectiveness and efficiency, of innovation and entrepreneurship activities in western Canada. The chapters provide a useful mix of subjects and methodologies. Chapter 6 offers a theoretical explanation, and remedy, for the lack of entrepreneurship and innovation within aboriginal communities. Chapter 7 presents an empirical study of the human resource practices of innovative firms. The final chapter of this section describes organizations designed to facilitate interactions among innovative entrepreneurial firms. These chapters continue to focus on western Canada, but the implications of their findings extend beyond the region.

Chapter 6: Developing market economies: The aboriginal case in northwest British Columbia

In the first chapter of this section, Ron Mitchell and Eric Morse introduce transaction cognition theory, a theory that integrates transaction cost economics with cognition theory. They apply the new theory to the economic development requirements of aboriginal peoples in western Canada. The authors explain four types of cognitions that make up their model. Planning cognitions are mental models that assist in developing analytical structures to solve problems. Promise cognitions are mental models that help promote trust in economic relationships. Competition cognitions are mental models that help create sustainable competitive advantages in economic exchanges, and political cognitions are the compliance, self-protection, and authoritarian cognitions that govern group actions. Mitchell and Morse use this powerful theoretical approach to explain the successes and breakdowns in attempts at entrepreneurship among aboriginals and offer insights as to how the strong traditions and culture of aboriginal communities might be directed toward productive ends.

There is still work to be done in fully operationalizing transaction cognition theory. However, it offers considerable promise in explaining how macro- and micro-economic systems operate. Specific to the focus of this book, the theory is a powerful new way of looking at entrepreneurship, family business, and economic development, and has potential for explaining the workings of innovation systems at the regional, national, and international levels.

Chapter 7: The link between innovation and the use of human resources in British Columbian enterprises

Adam Holbrook and Lindsay Hughes provide an interesting comparison of the human resource practices of innovative and non-innovative firms in British Columbia using two definitions of innovation: "new to the firm" and "new to the market." Although determining what is new to the market is a more difficult proposition, Holbrook and Hughes' study suggests that the additional insights such a distinction offers may make the effort worthwhile. Their results show that human resource dimensions such as post-secondary education of workforce, training programs, and availability of qualified personnel are no different for innovative and non-innovative organizations when innovation is defined as new to the firm. When innovation is defined as new to the market, however, the same dimensions effectively discriminate between innovative and non-innovative firms.

Based on their findings, Holbrook and Hughes argue that the "new to the market" definition should be used, and their results suggest that human resource practices may be an important indicator for distinguishing innovative and non-innovative firms. The authors also argue that the importing and exporting of knowledge may be a good indicator of the expertise of a regional system of innovation, and hence, a measure of development. We see this as a useful insight and supportive of the work of Phillips (Chapter 2) and Chiasson (Chapter 5). Holbrook and Hughes also call for additional research as to how innovation systems are developed and operated. Fortunately, with the recent advances in the mapping of such systems by Smith (Chapter 3), we now have a tool that will facilitate that task.

Chapter 8: Networking support to innovative young firms: Pathways to knowledge

The final chapter in the second section deals with network support for innovative young firms. In this chapter, Cooper Langford, Anne Tyrie, and Janet Peace draw on Ebers (1997) to argue that activity networks, trust, and catalysts are essential elements of innovation networks. They then expand on the contribution of Josty (Chapter 4) by exploring in depth a number of catalytic agencies that promote networks of relationships among innovators in Alberta. Their discussions are based on interviews with key participants in these organizations designed to investigate six propositions: (1) Subject Matter Experts use networking as a source of new knowledge and will increasingly do so; (2) Networking sustains innovation; (3) Networks are multidimensional relationships; (4) Several types of organizations sustain these networks; (5) The benefits of successful networking must outweigh the costs; and (6) Network brokers are valuable but rare. Their findings cast further light on the propositions and show that local and non-local support organizations and universities can be helpful in network formation.

Section III:
Innovations in entrepreneurial support programs

The third section of the book consists of three chapters dealing with entrepreneurial education and support programs. Whereas Chapter 8 by Cooper, Tyrie, and Peace deals with catalysts, the chapters in this section deal with the more direct impacts made possible by effective entrepreneurial support programs. Two (Chapters 10 and 11) describe programs in the Faculty of Management at the University of Calgary. The lessons learned from their successes and failures will be instructive to anyone interested in initiating or improving entrepreneurship education or training programs in their regions. On the other hand, Chapter 9 is different from the most of others in this volume in that it deals with innovations in entrepreneurial support programs that have largely been developed and implemented outside western Canada. In that chapter the authors, two of whom reside in western Canada, argue that these innovations can and should be adopted in the local region.

Chapter 9:
Lessons from successful innovations in entrepreneurial support programming

Ed McMullan, Jim Chrisman, and Karl Vesper identify four successful methods of delivering entrepreneurial education and training. The U.S. Small Business Development Center (SBDC) program has achieved its success in assisting pre-venture entrepreneurs and small business owners by counselling rather than consulting, with clients performing the work themselves under the direction of trained counsellors. By way of contrast, the Fastrac and Western Entrepreneurial Network programs in the U.S. offer formal 13-week courses for such clients. Unlike the SBDC, which is most effective when counselling ventures with the potential to become substantial enterprises, these two programs appear to be most effective in counselling micro-enterprises and home-based businesses. The Swinburne program in Australia has achieved a phenomenal success rate in its certificate (one-year), diploma (two-year), and degree (three-year) programs aimed at graduate students interested in becoming entrepreneurs. Finally, the Small Business Institute (SBI) in the U.S. and the Projects Program at the University of Calgary achieved success by providing students a form of on-the-job training, working with local entrepreneurs to develop business plans and conduct market research.

The authors argue that each of these successful innovations should be implemented and/or extended in the western Canadian context, and they identify a number of common attributes that can guide further development. Thus, a strategic focus, the development of tacit knowledge, linking education to action-based outcomes, just-in-time delivery, and client self-selection all appear to be important. While future research is needed, their research suggests that entrepreneurship can be "taught" and that effective entrepreneurship education programs are transferable to the western Canadian context.

Chapter 10: Developing entrepreneurs in a hybrid management and entrepreneurship MBA: A case study in Calgary

Whereas Chapter 9 dealt with successful innovations, Chapter 10 by Ed McMullan and Vance Gough describes a program innovation at the University of Calgary that was not an unqualified success. In 1993, Calgary changed its traditional MBA program into an Enterprise MBA program with a hybrid entrepreneurship/management focus. Despite the problems involved in balancing the education needs of middle managers and entrepreneurs, the high level of commitment of the core group who designed the program produced significant initial success in facilitating entrepreneurship. Changes in instructors resulted in uneven levels of commitment. This, along with the continuing struggle to reconcile the different needs of middle managers and entrepreneurs, led the program to drift back toward a traditional MBA over time.

McMullan and Gough argue that despite the successes and publicity that the Enterprise MBA garnered, effective entrepreneurship education cannot be sustained with half-measures or half-hearted commitments. In fact, in the intervening period between the preparation of this chapter and its publication, the University of Calgary has abandoned the program. An implication of McMullan and Gough's study is that entrepreneurship is not management, or at least, is a sort of management that is difficult to reconcile with traditional management education. Given the experiences of a recognized international leader in entrepreneurship education, their results call into question both the manner in which entrepreneurship is taught in North American universities and whether entrepreneurship education can be effectively taught in schools of business with traditional orientations geared to developing managers for large corporations.

Chapter 11: A preliminary assessment of the effectiveness of project-based assistance to medical entrepreneurs at the University of Calgary

In the final chapter of this section, Myron Weber, Jim Chrisman, Jim O'Grady, and Vance Gough share the results of another Calgary experiment in entrepreneurship education, the Student Projects Program of the Entrepreneurship and Innovation Area of the University of Calgary's Faculty of Management. In examining this successful project-based mode of education, the authors attempted to determine whether medical entrepreneurs could gain valuable assistance from MBA students in commercializing their discoveries and inventions. They suggest that project-based assistance is particularly well suited to the temperaments and needs of medical entrepreneurs, and the case studies indicate some support for that proposition. The authors conclude that further efforts to match medical entrepreneurs, long on technical expertise and short on business acumen, with supervised groups of graduate students in entrepreneurship, might yield considerable educational benefits for both groups.

Although the results of the study show that there are informational benefits to be gained by medical entrepreneurs from the work of student teams, the program may not be as beneficial in facilitating the development of the tacit knowledge that McMullan, Chrisman, and Vesper (Chapter 9; see also Chrisman and McMullan 2000) suggest is crucial for successful entrepreneurial support programs. Rather, in our view the primary benefit of the project program may be its potential for unfreezing the attitudes of medical researchers toward outsider assistance. Given the culture clashes between medical researchers and business people noted in the article, the value of gradually developing an entrepreneurial mindset among medical researchers should not be underestimated.

Section IV: Innovation, entrepreneurship, and management in family businesses

The final section of this book consists of five chapters devoted to family business management. As noted previously, family businesses frequently evolve from successful innovative and entrepreneurial firms. If these businesses are to realize their economic development potential, they must continually renew their competitive advantages. This requires mobilization of financial, organizational, human, and relational resources.

The five chapters in this section deal with different aspects of these resources. Chapter 12 is about securing venture capital for the family business, while Chapter 13 discusses the requisites for preserving the spirit of innovation within the family firm. Chapters 14 and 15 are concerned with human resources, the former covering the importance of non-family managers to the growing family business and the latter dealing with executive compensation – a major motivator for risk-taking. Chapter 16 explores trust and its effective limits as a resource when a business becomes more and more complex. More than any other section, this group of chapters concerns the human element of entrepreneurship and innovation. Although only Chapter 12 deals specifically with the western Canadian context, all five chapters offer considerable insights into the strengths and issues of family firms in the region. The perspectives of these western Canadian authors enrich the study of family businesses in Canada in general. Finally, and just as importantly, these chapters will hopefully draw the attention of researchers and practitioners to this important and oft-neglected form of organization.

Chapter 12: Investing with families in business: A venture capitalist's view

Robert Slator's speech, edited by Jess Chua, provides an interesting perspective on the potential of family businesses from an investor's point of view. Slator states that venture capitalists outside western Canada do not invest in family businesses because most family firms are not the high-potential firms that justify the high-cost, high-involvement capital supplied by venture capitalists. Within western Canada he estimates that about 50% of his company's investments in family businesses have been successful. However, he also notes that significant changes in the "family compact" had to be made in many of those cases. Slator suggests that regardless of its form of ownership, a successful, high-potential venture must have a strong entrepreneur, market leadership, a global perspective, and appropriate financing.

In addition, family businesses must resolve certain issues unique to the family firm. These include deciding early whether the firm's objective is market leadership or maintaining lifestyle, keeping family member rivalry under control, steering clear of nepotism, and clarifying the leadership hierarchy. He concludes by noting that if such issues can be resolved, a family business will have greater cohesiveness, a stronger common purpose, and potentially, greater likelihood of obtaining significant external capital.

In the global economy, no business is shielded from global competition. Therefore, family businesses will need to tap increasingly large amounts of external capital just to maintain position, let alone be the market leaders envisioned by Slator. The observations made in this chapter are not hypotheses or propositions generated by rigorous qualitative research, but they can form a basis for further conceptual development and empirical research on the attributes that a family firm must possess in order to obtain venture capital financing.

Chapter 13: Your old men will dream dreams, your young men will see visions: A conceptualization of innovation in family firms

There is an extensive literature on innovation in organizations and innovative organizations. How this may be different in a family business has not been frequently explored in the literature. The chapter by Reg Litz and Robert Kleysen represents an initial attempt to fill this gap. The authors use a variety of case examples, including the innovative activity of the Brubeck family in jazz improvisation, to explore intergenerational innovation – innovation across generations – in family firms. They identify four important elements of the intergenerational innovation process in the family firm. First is the deliberate effort by the parents to facilitate the development of requisite knowledge and skills in the next generation. Second is parental relinquishment – willingness on the part of the parents to let the children be their own special individuals. Third is the formation of a new customer value proposition by the next generation. Fourth is a flexible definition of success.

These interesting observations point to many avenues for both theoretical and empirical research. What is innovation for the family firm? For example, is a family firm that diversifies its business through acquisition innovating? Do family firms have relational advantages in innovation? Are these advantages for technical, organizational, managerial, transactional, or financial innovations? What role does gender, culture, or ethnicity play? Do intragenerational and intergenerational innovations follow the same process?

Chapter 14: Family and non-family priorities in family firms: Preliminary evidence

Jess Chua, Jim Chrisman, and Pramodita Sharma note that as the family firm expands, its dependence on non-family managers grows. Using data from a large sample of family businesses in Canada, the authors determine that dealing with non-family managers is considered by family firms to be a very important concern. The data also show that the concern tends to increase with size, the number of family business managers, and the likelihood of non-family succession.

These results suggest that family firms realize they cannot continue to grow and excel without loyal, committed, and involved non-family managers. Chua, Chrisman, and Sharma note that, despite the high level of this concern, little research has been done with respect to non-family managers in the family business. The paper should motive researchers to explore how such loyalty, commitment, and involvement may be secured so that firms can continue to be innovative and entrepreneurial past the initial stages of development and the first generation of the founding family.

Chapter 15: Executive pay practices in family-controlled firms in Canada: A preliminary study

Compensation is clearly a very important motivator of both family and non-family managers. Yun Park's study compares and contrasts the executive pay practices of family and non-family firms in Canada using the insights of agency theory. Agency theory within this context is concerned with the conflict of interest between shareholders and managers if they are not the same individuals (Alchian and Demsetz 1972; Jensen and Meckling 1976). This conflict of interest can be controlled using close monitoring by shareholders or by adopting appropriate incentive compensation for the managers. Park hypothesizes that family firms have a cost advantage in monitoring management. Therefore, since share-based incentives are more costly, family firms are less likely to rely on these incentives to resolve agency problems. He also hypothesizes that since executive stock options lead to higher pay because of inefficient risk sharing, the overall reported executive pay in family firms will be lower than that in non-family firms. Park's findings support both hypotheses. These results contradict those from other studies that find pay packages for non-family managers in family firms to be higher because the non-family managers must accept the fact that they have little or no chance of ever being promoted to the very top of the company (Poza, Alfred, and Maheshawi, 1997).

Park's results are based on publicly traded firms. In his study, family firms are those with an individual or family owning 20% or more of the company. It will be interesting for future researchers to see whether his findings also apply to privately-held family firms. Whether family firms are in the sibling partnership or cousin consortium generation (Gersick, Davis, Hampton, and Lansberg, 1997) may also make a difference. Finally, agency theory assumes that both shareholders and managers operate entirely on the basis of self-interest. There has recently arisen a group of researchers who believe that family firms do not work this way. Instead, they believe that family business managers, especially those who are family members, act more like stewards, putting the interest of the firm above their personal interests (Zahra, 2001).

Chapter 16: Exploring the role of trust in family firm governance

Trust is the relational resource most frequently cited anecdotally as the driver of a family firm's competitive advantage. In the last chapter of this book, Lloyd Steier discusses trust within family firms, a subject that has not received much attention in the family business literature. Organizational theorists have observed that trust has its limits despite its contribution toward a cost advantage through lower monitoring costs (Wicks, Berman, and Jones 1999). For example, fraud arises often as the result of excessive trust. How to achieve an appropriate balance is a dilemma all firms must resolve. Steier applies this concept to the family firm and observes that this balance will shift as the family business evolves. For example, as the family firm grows through the natural stages of the founder-manager, sibling partnership, and cousin consortium (Gersick, Davis, Hampton, and Lansberg, 1997), trust may decrease with the complexity of the relationships among family members. Consequently, organizational structure and policies may have to shift away from reliance on trust and become more formalized.

Considered with the other chapters in this section, trust appears to influence how executives are compensated, the relationships between family and non-family members, and the instigation and success of innovative efforts. Will distrust inevitably increase with relational complexity? Are there ways that the family firm can preserve trust through family ties and bonding? These are important questions for family business managers because they will affect the family firm's cost structure, its innovativeness, and its competitiveness.

Integration

The subjects examined in this book could be organized along many different lines. For example, they could be separated into those dealing with how innovation occurs and those concerned with how to facilitate innovation. Alternately, the articles could be organized into those dealing with the networks, people, and social capital required for successful innovation. The chapters also cover a wide range of methodologies. These include purely conceptual pieces, theory-building exercises based on field research, literature surveys, the practical state-of-the-art, large-scale empirical studies, and presentations of exciting new methodological advancements that will aid research in the future. The mix in topics and methodologies does not really permit a clear separation. Consequently, our approach strives for similarities among subjects within a section and differences across sections. At the same time, we believe that several themes link the chapters within and across the sections of the book.

There are obviously many factors that affect the level of innovation and entrepreneurship, and incidence of family business, within a region or nation. Three themes – knowledge, trust, and support systems – persist in the studies included here. The collective arguments and evidence presented in this book lead us to conclude that the exchange of ideas and knowledge will be effective and efficient only in the presence of the appropriate mix of people, trust, and effective support systems.

Knowledge

Central to the envisioned process of entrepreneurship and innovation within an innovation system is the exchange of ideas or knowledge among people. Several chapters in this book are directly or indirectly concerned with how and among whom knowledge is exchanged. Phillips (Chapter 2) studies how ideas are imported into the Saskatoon biotechnology cluster, improved, and exported. Josty (Chapter 4) investigates the main players in the Alberta innovation system. Smith (Chapter 3) uses techmaps to trace the transfer of technology and people among high technology firms. Langford, Tyrie, and Peace (Chapter 8) look at networking support as the pathway of knowledge. Mitchell and Morse (Chapter 6) discuss the exchanges that occur among aboriginal peoples and governments. Finally, Litz and Kleysen (Chapter 13) illustrate quite convincingly how exchanges of information between family members may have a profound influence on whether and how much innovation occurs in such organizations.

Other chapters explore how this exchange may be facilitated. Chiasson (Chapter 5) proposes a compromise within the public policy dilemmas encountered when managing the generation and application of ideas in furtherance of economic development. McMullan, Chrisman, and Vesper (Chapter 9) discuss the general characteristics of successful innovations in entrepreneurial support programming.

Education is one of the most important channels of knowledge exchange. Two chapters look at the effectiveness of innovative educational programs in promoting entrepreneurship. Weber, Chrisman, O'Grady, and Gough (Chapter 11) examine project-based assistance to medical entrepreneurs while McMullan and Gough (Chapter 10) look at a hybrid MBA program. In other chapters, Holbrook and Hughes (Chapter 7) examine how human resources may be managed to encourage innovation; Slator (Chapter 12) discusses how participants in family businesses must communicate with each other and investors in order to achieve their economic ends.

Trust

While people are the generators and users of ideas and knowledge, exchange will not happen unless there is trust, even if the exchange is based on a contractual commercial agreement. Bounded rationality precludes the design of complete contracts – contracts that cover every contingency. Steier (Chapter 16) is explicitly concerned with the issue of trust as it applies to family firms. However, the importance of trust permeates many of the other chapters.

Thus, the very term "cooperative mercantilism" that Chiasson (Chapter 5) uses to describe how regional systems of innovation should operate presupposes that trust will exist among organizations and governments within and across regions. Indeed, Phillips' (Chapter 2) description of the Saskatoon canola industry suggests a level of cooperation and information exchange that could not exist without some degree of trust between the various players. Mitchell and Morse (Chapter 6), in discussing economic development among aboriginal peoples in western Canada, make it clear that a key component of their model is trust and that a key factor in explaining the problems those communities have had is a lack thereof. In the setting of family firms, Park (Chapter 15) shows that trust or its absence has a major impact on a firm's compensation decisions, while Litz and Kleysen (Chapter 13) argue that trust may determine whether or not the firm engages in innovation. Chua, Chrisman, and Sharma (Chapter 14) remind us that trust between family and non-family managers is a critical concern of the former and needs to be added to our research agenda.

Support systems

The last theme on which a number of chapters touch is the need for support systems in promoting innovation and entrepreneurship. These institutions may be public, quasi-public, or private. Chiasson (Chapter 5), Mitchell and Morse (Chapter 6), Josty (Chapter 4), and Langford, Tyrie, and Peace (Chapter 8) emphasize the positive, or negative, role governmental bodies can play. Weber, Chrisman, O'Grady, and Gough (Chapter 11) and McMullan and Gough (Chapter 10) examine education, typically a publicly supported activity, as a fostering process in entrepreneurship and innovation. McMullan, Chrisman, and Vesper (Chapter 9), on the other hand, show that private foundations have picked up an important part of the load. Finally, Slator (Chapter 12) shows that, in western Canada at least, venture capital funding and venture capitalist mentoring are important parts of the support system for family businesses that wish to pursue entrepreneurial opportunities with high growth potential.

Conclusions

This introductory chapter acquaints the reader with the wide range of diverse but interrelated topics covered in the book and provides an overview of the individual chapters. It also identifies some of the themes that cut across the contributions to the literature of the various chapters. Undoubtedly, others will notice other commonalities and differences. We will consider our objective realized if the reader finds value among the chapters included herein. We also hope that the reader will come away with a new perspective on, and appreciation of, the topics of innovation, entrepreneurship, and family business as well as a new understanding of their contribution to the development of the economies in western Canadian, Canada in general, and regions and nations in other parts of the world.

References

Acs, Z.J., and Audretsch, D.B. (1990). *Innovation and Small Firms.* Cambridge: MIT Press.

Alchian, A, and Demsetz, H. (1972). Production, information costs, economic organization. *American Economic Review* 62: 777-95.

Beckhard, R., and Dyer, W.G., Jr. (1983). Managing continuity in the family owned business. *Organizational Dynamics* 12(1): 5-12.

Birch, D. (1987). *Job Creation in America*. New York: Free Press.

Burgelman, R.A., and Sayles, L.R. (1986). *Inside Corporate Innovation: Strategy, Structure, and Managerial Skills*. New York: Free Press.

Bygrave, W.D., and Hofer, C.W. (1991). Theorizing about entrepreneurship. *Entrepreneurship Theory and Practice* 16(2): 13-22.

Chrisman, J.J., Bauerschmidt, A., and Hofer, C.W. (1998). The determinants of new venture performance: An extended model. *Entrepreneurship Theory and Practice* 23(1): 5-29.

Chrisman, J.J., and McMullan, W.E. (2000). A preliminary assessment of outsider assistance as a knowledge resource: The longer-term impact of new venture counselling. *Entrepreneurship Theory and Practice* 24(3): 37-53.

Chua, J.H., Chrisman, J.J., and Sharma, P. (1999). Defining the family business by behaviour. *Entrepreneurship Theory and Practice* 23(4): 19-39.

Deloitte & Touche (1999). *Are Canadian family businesses an endangered species?* Waterloo: University of Waterloo and Deloitte & Touche Centre for Tax Education and Research.

Ebers, M. (1997). *The Formation of Inter-Organizational Networks*. Oxford: Oxford University Press.

Gartner, W.B. (1988). "Who is an entrepreneur?" is the wrong question. *American Journal of Small Business* 12(4): 11-32.

Gersick, K.E., Davis, J.A., Hampton, M.M., and Lansberg, I. (1997). *Generation to Generation: Life Cycles of the Family Business*. Boston: Harvard Business School Press.

Herron, L., and Robinson, R.B., Jr. (1993). A structural model of the effects of entrepreneurial characteristics on venture performance. *Journal of Business Venturing* 8: 281-94.

Jensen, M.C., and Meckling, W.H. (1976). Theory of the firm: Managerial behavior, agency costs and ownership structure. *Journal of Financial Economics* 4: 305-60.

Poza, E.J., Alfred, T., and Maheshwari, A. (1997). Stakeholder perceptions of culture and management practices in family and family firms – A preliminary report. *Family Business Review* 10: 135-55.

Reese. L., and Fasenfast, D. (1997). What works best? Values and the evaluation of local economic development policy. *Economic Development Quarterly* 11: 195-207.

Rogers, E.M. (1962). *Diffusion of Innovation.* New York: Free Press.

Schumpeter, J.A. (1934). *The Theory of Economic Development.* New Brunswick, New Jersey: Transaction Publishers.

Shanker, M.C., and Astrachan, J.H. (1996). Myths and realities: Family businesses' contribution to the U.S. economy. *Family Business Review* 11: 107-23.

Sharma, P., and Chrisman, J.J. (1999). Toward a reconciliation of the definitional issues in the field of corporate entrepreneurship. *Entrepreneurship Theory and Practice* 23(3): 11-27.

Sharma, P., Chrisman, J.J. and Chua, J.H. (1997). Strategic management of the family business: Past research and future challenges. *Family Business Review* 10: 1-35.

Shleifer, A., and Visny, R. (1997). A survey of corporate governance. *Journal of Finance* 52: 737-83.

Soukup, W.R. (1979). *Strategic responses to technological threat in the electronic components industry.* PhD dissertation, Purdue University.

Tan, W.L., and Fock, S.T. (2001). Coping with growth transitions: The case of Chinese family businesses in Singapore. *Family Business Review* 14: 123-39.

Utterback, J.M. (1994). *Mastering the Dynamics of Innovation.* Boston: Harvard Business School Press.

Von Hippel, E. (1988). *The Sources of Innovation.* New York: The Oxford Press.

Wicks, A.C., Berman, S.L., and Jones, T.M. (1999). The structure of optimal trust: Moral and strategic implications. *Academy of Management Review* 24: 99-116.

Williamson, O.E. (1985). *The Economic Institutions of Capitalism: Firms, Markets, Relational Contracting.* New York: Free Press.

Zahra, S.A. (2001). Ownership and involvement and international expansion : An empirical test of the stewardship theory among family firms. Paper presented at Theories of Family Enterprise: Establishing a Paradigm for the Field. First Annual Conference of Enterprising Families, Edmonton.

I

Innovation Systems

Chapter 2

Regional Systems of Innovation as Modern R&D Entrepots: The Case of the Saskatoon Biotechnology Cluster

Peter W.B. Phillips

Absract

Economics examines innovation from the perspective of the incentives for, and impacts of, research at the microeconomic or firm level, while the national systems of innovation (NSI) literature examines clusters of innovation within states or regions. This paper offers a synthesis of the two approaches to studying innovation, proposing a unified theory that posits that systemic modern innovation occurs within and among firms and agencies at the local or regional level while integrated into the global production system. As such, modern innovation systems reflect many of the characteristics of the classical trade entrepot, where most inputs are imported tax-free, value is added locally, and semi-finished outputs are exported for further processing and distribution to final consumers. The Saskatoon-centred, canola-based biotechnology cluster exhibits many of these attributes. The majority of the basic research and many of the proprietary technologies are imported, assembled into new crop varieties for commercial release in western Canada first, and then exported on a largely tax-free basis as intermediate product to global markets. This alternative model for innovation poses significant implications for researchers, markets and public policies.

Introduction

Innovation, the driver for growth and development, is increasingly being discussed and examined. There are two main competing views of innovation. The economic theory of technological change has, for many years, focused on the firm as the primary research unit (Solow, 1956; Arrow, 1962) and, in the footsteps of Schumpeter (1954), has examined the microeconomic incentives and impacts of private research for commercialization. Some economists (Romer, 1990; Grossman and Helpman, 1991; Krugman, 1998) recently have examined the impact of innovative activity at the firm level on the larger economy, focusing especially on the implications of endogenously generated innovation on macroeconomic growth, trade and industrial location. Even so, they tend to assume that innovation is a discrete event that occurs within firms. Recently, a few of these economists have begun to notice that firms are not complete and must often reach beyond their boundaries to satisfy their needs. Nevertheless, the focus remains steadfastly on innovation that is primarily directed by a single corporate leader.

Alternatively, a group of political economists with more of an interest in the influence of institutions has developed a theoretical systems approach to innovation – the national systems of innovation (NSI). Harkening back to Marshall (1890), these researchers (e.g., Freeman 1987; Nelson 1988; Porter 1990; Lundvall 1992) have looked at the role of economies of scale and scope in the local, regional, or national innovation systems themselves. Once again, however, this approach tends to consider innovation as an activity that is limited, in this case to a region. While the NSI literature posits that innovation is embedded in networks that at times span international borders, the resulting policy prescriptions tend to focus on how national governments can encourage greater self-sufficiency in innovation. Those that have looked at the international effects have examined how multinational enterprises (MNEs) contributed to domestic capacity through direct investment (Chesnais 1992) and at innovation's effects on international trade (Dalum 1992; Andersen and Broendgaard 1992; Fagerberg 1992).

In practice, standing back from either the firm or the region and looking instead at the sources and uses of knowledge within the innovation system provides a significantly different perspective. This paper, drawing on evidence in the agricultural research world, presents a synthesis approach that starts from the assumption that innovation is actually a global activity that transcends any firm or region. No one firm or region drives the innovative activity – multiple actors jointly lead it. Parts of the innovative effort, however, have the potential to become linked to firms or regions. In essence, modern innovation systems reflect many of the characteristics of the classical trade entrepot, where most of the inputs are imported, value is added locally, and semi-finished outputs are exported for further processing and distribution to final consumers.

The Saskatoon-centred, canola-based biotechnology cluster exhibits many of these attributes: the majority of the basic research and many of the proprietary technologies are developed elsewhere and imported; they are assembled into new crop varieties for commercial release first in western Canada; then they are exported on a largely tax-free basis as germplasm or intermediate product to global markets. Consistent with both the economic and NSI theories, canola research is agglomerating in and around Saskatoon, posing significant implications for researchers, markets, and public policies.

This chapter examines the theory, practice, and implications of regional systems of innovation as entrepots. The second section offers a short outline of the background to the regional innovation system operating in Saskatoon. The third section provides a summary of the theory and literature on clusters and innovation systems; the fourth section briefly outlines the data sources used to examine the Saskatoon cluster. The fifth section examines the Saskatoon cluster, highlighting the entrepot aspects of the innovation system, and the last section provides a discussion of some of the policy implications of this alternate approach to innovation research.

Background and Circumstances

"Location, location, location," the battle cry for property realtors everywhere, is increasingly becoming the focal point for discussion of the dynamics and benefits of knowledge-based growth. This examination of the Saskatoon-centred, canola-based research cluster can help to illustrate the dimensions and structures that underlie a successful knowledge-based research system.

The original transformation of rapeseed into canola, and more recently the privately-directed, research-intensive activity around canola, is largely a Canadian story (McLeod 1974; NRC 1992; Gray, Malla, and Phillips 1999; Phillips and Khachatourians 2001). For thousands of years, rapeseed has been used in various markets as a cooking oil, industrial lubricant, animal feed or green manure. After 1943, focused research in universities, at Agriculture Canada (now Agriculture and Agri-Food Canada, AAFC) and at National Research Council (NRC) labs in Canada sought to improve the agronomic and food attributes of rapeseed. That effort culminated in 1978 with the development of a new variety of rape that had lower amounts of erucic acid (linked in some studies to heart problems) and glucosinolates (which impair the feed value of the meal). The new low-erucic acid, low-glucosinolate standard was trademarked as "canola" by an industry association and became the forerunner of most varieties used around the world today. Beginning in the 1980s with the adoption of genetic engineering technologies, canola was targeted for further development by private companies. In 1995, canola became the first genetically-modified food

Table 1: Canada's and Saskatchewan's share of the total canola research effort globally, and proportion of share produced by public sector

	Global employment		Saskatchewan		Canada	
	Annual average	Public effort as % total	Share of global total	Public effort as % share	Share of global total	Public effort as % share
80-84	191	68.3%	32.2%	96.8%	48.1%	94.6%
85-89	287	56.1%	25.8%	96.8%	42.1%	85.3%
90-94	499	44.1%	25.3%	81.3%	41.3%	65.9%
95-98	702	36.4%	29.5%	59.5%	48.3%	44.5%

Source: Phillips and Khachatourians, 2001; survey of canola industry undertaken in 1997-8; based on person-years of employment devoted to research and development

crop in the world to be extensively commercialized, and it now has one of the highest adoption rates of any genetically modified food crop. Research and development into new novel traits continues.

A survey of most of the companies and public sector research and development efforts in the canola industry (Phillips and Khachatourians 2001) revealed that since the advent of private capital in the sector in the mid-1980s, the canola research effort has been slowly but steadily concentrating in Canada, specifically in Saskatoon. Table 1 shows that Canada's share of the total global research effort, which was approximately 75% in the 1945-66 period, dropped to about 40% in the late 1980s as the research effort disseminated from the public to the private sector. It has since rebounded to approximately half of the global effort. This concentration has been almost exclusively due to the relocation or development of private research and development programs in Canada. The public share of the Canadian effort has dropped from almost 100% in the 1944-75 period to about 44% in the late 1990s. Meanwhile, the effort in Canada has concentrated in Saskatoon. Although companies like Zeneca, Pioneer Hi-Bred, and Cargill/Intermountain Canola have located in other provinces, extensive private research has located to Saskatoon. Private research is supplemented by an increasing concentration of public research by Agricultural and Agri-Food Canada and the National Research Council in the city since the late 1980s.

Preliminary evidence suggests that the vast majority of the applied research in the development of canola varieties is based in Canada. Canada has registered 275 varieties of rapeseed/canola since the 1940s, more than two-thirds developed domestically. Although some other countries (especially China, Sweden, and Germany) developed varieties, all of the new traits introduced within the past 40 years (e.g., low erucic acid, low glucosinolates, hybrids, novel oils, blackleg disease resistance, key yield gains) were introduced first and at least partly developed in Canada. As a result, in 1999 western Canadian farmers produced more than 20% of the world's output of canola/rapeseed, exporting approximately 80% of it to global markets, accounting for more than half of the world trade (Phillips and Khachatourians, 2001).

From a superficial perspective, this level of agglomeration suggests that Saskatoon and Canada may have reached a critical, self-sustaining mass of capacity. In practice, however, this agglomeration has limited scope. Saskatoon and Canada account for a much higher share of some parts of the research, development, and commercialization of new innovations than they do of other parts. Saskatoon imports much of the basic research, patented technologies, and skilled workers, combines them with locally-owned and developed elite germplasm, commercializes the new varieties in western Canada and then markets the raw or semi-finished product to the global agri-food industry. By conservative estimates, however, less than half of the end-market value of canola produced in Canada is added in Canada; the rest is either imported and added during the R&D or production phases or added as the seed, oil, and meal moves downstream toward the consumer (Phillips and Khachatourians 2001). In short, Saskatoon is a centre for innovation, but cannot be viewed as self-sustaining. Rather, it exhibits traits that hark back to the entrepot model of development.

The Model

There is significant confusion in the literature about the definition, sources, and uses of innovation. For the purposes of this paper, it is assumed that innovation involves planned, systematic, sustained efforts to add value through research and development. In that sense, it differs from much of the invention that occurs in small and medium-sized companies, where individual entrepreneurs or scientists have eurekas that lead to valuable science or technologies. If the development is solely based on inspiration, neither the economics nor the systems literature has much to add. These developments are often better understood as the result of serendipity rather than planning and effort. Modelling this type of activity is pointless, as inspiration seldom arrives in the same place twice. In contrast, planned and systematic innovation can and should be modelled and examined. This section offers the historical entrepot framework as a new approach for evaluating the incentives and impacts of innovation on firms and regions.

The concept of entrepot trade has been largely expunged from current economic literature and can only be found in dated references. A quick review of the leading intermediate and advanced textbooks and reference books on international trade and development failed to unearth any reference to entrepot trade. It is necessary to go back more than a quarter century, to a time when the focus in trade studies was exclusively on goods, to find any reference to entrepots. At that time an entrepot was defined as "a centre at which goods are received for subsequent distribution. An entrepot port has facilities for the transhipment of imported goods or their storage prior to their re-export, without the need to pass through customs control" (Bannock, Baxter, and Rees 1972). Endacott's (1964) examination of Hong Kong, the classical example of a traditional entrepot, illustrates some of the key features of entrepots.

> The essential feature of the entrepot trade of Hong Kong was the existence of entrepot services, which tended automatically to attract trade.... [I]n addition to its natural harbour, Hong Kong possesses assets that were almost equally potent in making it a successful commercial centre. It possessed among its people, business acumen, managerial ability, commercial experience, professional skill, financial resources, control of shipping and a good supply of industrious and inexpensive artisans and workers. It would be an exaggeration to say that the entrepot trade was a product of these entrepot services ... but ... the rise of Hong Kong rested on the twin pillars of shipping and commercial skill ... functioning under the security of a British administration (1964, xiii).

One can see that even with this early model of entrepot trade, value-added services were a critical part of the success of these centres. Saskatchewan exhibits many of the attributes of this classical entrepot trade model. While Hong Kong relied on a natural harbour and shipping and Saskatchewan relies on an abundance of high quality agricultural land, both are fundamentally driven by the quality of their institutions and their people.

The challenge is to add the innovation dimension to the entrepot model. If new developments were the result of inspiration, then the traditional linear view of R&D would be appropriate. On the other hand, if one looks at systemic innovation processes, particularly at the many inputs and outputs they usually involve, it becomes clear that no single firm or region can truly be viewed as self-sufficient or self-sustaining.

Figure 1: The chain link model of innovation

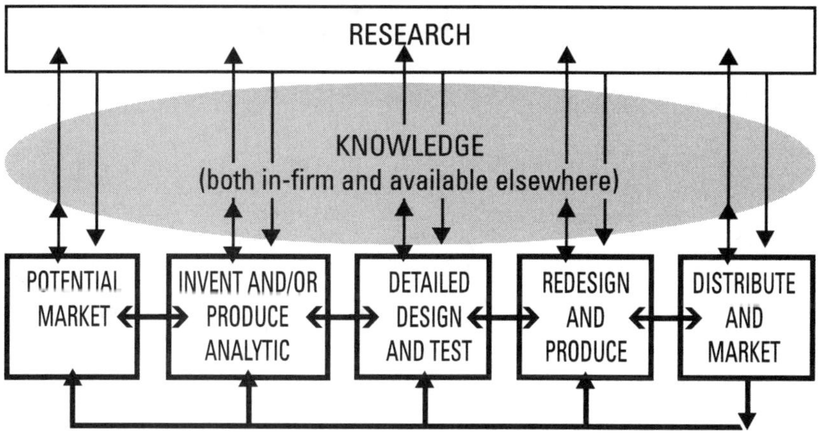

Klein and Rosenberg (1986) provide a non-linear approach that explicitly identifies the role of both market and research knowledge and the potential for open research systems. Their chain-link model of innovation (Figure 1) begins with a basically linear process moving from potential market through invention, design, adaptation, and adoption. However, it adds feedback loops from each stage to previous stages and the potential for the innovator to seek out existing knowledge or to undertake or commission research to solve problems in the innovation process. This dynamic model raises a number of questions about the types and roles of knowledge in the process. Some of the knowledge will be available or could be developed either within or outside the firm.

Malecki (1997) provides a way of categorizing types of knowledge that helps to identify which route a firm or institution might use to acquire or develop knowledge needed to innovate (Table 2). He identifies four distinct types of knowledge: know-why, know-what, know-how and know-who.

Table 2: Classification of types of knowledge

	Degree of Codification	Produced by	Extent of disclosure
Know-why	Completely codified	Universities and public labs	Fully disclosed and published in scientific journals
Know-what	Completely codified	Universities, public labs, and private companies	Fully disclosed in patents
Know-how	Not codified	Hands-on in labs	Tacit; limited dispersion
Know-who	Not codified	Exists within firms or research communities	Tacit; limited to community

Source: Adapted by author from Malecki, 1997, p. 58.

Each type of knowledge has specific features (OECD, 1996).

- "Know-why" refers to scientific knowledge of the principles and laws of nature, and for the most part is undertaken globally in publicly funded universities and not-for-profit research institutes. It is subsequently codified and published in academic or professional journals, making it fully accessible to all. This knowledge would be found in the knowledge block of the chain-link model, having been created almost exclusively in the research block.
- "Know-what" refers to knowledge about facts and techniques, which can usually be codified and transferred through the commercial marketplace. The stock of know-what is in the knowledge block of the chain-link model, having been created in the research, invention, design and adoption blocks.
- "Know-how" refers to the combination of intellectual, educational and physical dexterity, skills, and analytical capacity to design a hypothesis-driven protocol with a set of expected outcomes. It involves the ability of scientists to effectively combine the know-why and know-what to invent and innovate. This capacity is often learned through education and technical training and perfected by doing, which in part generates a degree of difficulty for the uninitiated and makes this type of knowledge more difficult to codify, and hence, to transfer to others. Know-how would be represented in the research block and also in the invention, design and adaptation stages of the chain-link model.
- Finally, "know-who," which "involves information about who knows what and who knows how to do what" (OECD 1996, 12), is becoming increasingly important in the biotechnology-based agri-food industry. As the breadth of knowledge required to innovate expands, it has become absolutely necessary to collaborate. In today's context, know-who also requires knowledge of, and access to, private sector knowledge generators who, at times, may hold back the flow of crucial and enabling information, expertise, and knowledge. Know-who knowledge is seldom codified but often accumulates within an organization or, at times, in communities where there is a cluster of public and private entities that are all engaged in the same type of research and

> development, and who often exchange technologies, biological materials and resources and pursue common staff training or cross-training opportunities. The arrows in the chain-link model would represent this type of knowledge, while the building of relationships that lead to trusting networks of know-who is the basis for the flows.

This chapter uses a modified chain-link model to investigate the Saskatoon-centred biotechnology cluster, in order to test the hypothesis that knowledge-based innovation clusters operate in ways similar to classical trade entrepots. Applying the conceptual structure of knowledge offered by Malecki (1997) and the chain-link innovation system proposed by Klein and Rosenberg (1986) to the innovation systems model developed by Lundvall (1992) allows an examination of the degree of self-sufficiency of either innovating firms or innovating regions. This framework is used to illustrate that innovation systems function to varying degrees as entrepots, depending on their stage of knowledge development and innovation.

The Data Sources

No definitive set of measures for knowledge and innovation has yet been developed. Nevertheless, there has been significant work undertaken in a number of areas that use proxies for knowledge and the transmission of knowledge. Using the four types of knowledge, and the resulting products, one can construct a package of empirical measures that approximate the flow of innovations through the research system and into the marketplace.

First, starting with know-why knowledge, it is clear that while it is quite difficult to identify the inputs to the research effort, one can look at bibliometric estimates to measure the flow of knowledge from creators, generally in universities, research institutes, and private firms. There is general acceptance of the view that publications such as academic journals are the primary method for communication of personal and institutional findings that become the vehicle for evaluation and recognition (Moed, Burger, Frankfort, and van Raan 1985). Hence, in general in the past and to some extent in current practices, most if not all of the effort put into research is ultimately presented for publication. The common catch-phrase, "publish or perish" captures the essence of past practice, while the more modern "patent and then publish" pattern exemplifies practices in a large number of research universities and public labs. There have been a number of efforts (e.g., Industry Commission 1995) to develop and use literature-based indicators to evaluate scientific effort.

For the purposes of this study, the Institute for Scientific Investigations (ISI) was contracted to undertake an electronic search of their databanks, covering 1981 to July 1996. They were instructed to search their database of approximately 8,000 journals in the sciences and social sciences for seven key words/phrases: brassica campestris, brassica napus, brassica rapa, canola, canola meal, rapeseed, and oilseed(s). The special tabulation identified 4,908 individual articles in 650 journals meeting the criteria (hereafter called the canola papers). They were produced by approximately 6,900 authors in approximately 1,500 organizations in 79 countries. The ISI data also provides the capacity to look both forward and backward from the target articles to determine where the key knowledge inputs come from and where the resulting knowledge is being used. The database identifies 17,995 papers that were cited. The cited papers were from 1,294 journals, produced by approximately 28,800 authors in 3,816 organizations in 107 countries which were cited a total of 28,946 times by the 4,908 papers that relate to canola research. At the other end of the system, the 4,908 canola papers were cited 26,946 times. The database can also be sorted and searched by author, institution, subject and country of the researcher, and then cross-tabulated for collaborations, allowing examination of both the stocks and flow of knowledge. In this way, one can investigate the know-who linkages that underpin the innovation system.

Second, know-what knowledge is most commonly examined using patent information. Trajtenberg (1990, 183) argues that "patents have long exerted a compelling attraction on economists dealing with technical change.... The reason is clear: patents are the one observable manifestation of inventive activity having a well-grounded claim for universality." Trajtenberg (1990, 215) concludes that in the context of specific, clearly demarcated innovation (in his case, CT scanners), patents "play an important role in studying the very emergence of new markets, which seems to be the period when most of the innovative activity takes place." He likens patents to working papers in economics. Papers and patents are produced roughly in proportion to effort: a larger number of papers or patents indicates a larger research effort. "Patent counts can thus be regarded as a more 'refined' measure of innovative activity than R&D, in the sense that they incorporate at last [sic] part of the difference in effort, and filter out the influence of luck in the first round of the innovative process" (Trajtenberg 1990, 216). For the purposes of this study, the Canadian Intellectual Property Office (CIPO) database of Canadian patent bibliographic data was searched for canola-related patents in late December 1999. That database contains all applications for patents made between 1920 and December 1999. During that period, 634 patents for canola-related work were filed.

Know-how and know-who knowledge, as discussed above, are often inseparable and are tricky to track at the best of times. Nevertheless, these types of knowledge can be mapped by looking at a number of different sources. The regulatory systems in Canada and elsewhere provide one means of identifying who is converting the know-why and know-what knowledge into actual products, as they assess risks during the detailed design, testing, and redesign periods. This data is available in Canada through the Canadian Food Inspection Agency (CFIA) authorizations for field trials of "plants with novel traits" and internationally through the Organization for Economic Cooperation and Development (OECD) website on field trials. Moving along through the innovation system, the resulting products can be observed through the varietal registration system in Canada under the Seeds Act and in Canada and elsewhere through the registration of new canola varieties for plant breeders' protection, as provided under the International Union for the Protection of New Varieties of Plants (UPOV). This data must be supplemented by industry data to identify public varieties that are not protected by breeders' rights.

Finally, the ultimate measure of innovative success is commercialization and market adoption. Information on capacity in related and supporting industries is available through industry surveys, but data on the seeds market is becoming more difficult to find. Aggregate data for canola acreage and yields are available nationally and through the Food and Agriculture Organization (FAO), but production information on specific varieties is difficult to obtain. The canola industry in Canada is fortunate to have relatively good public sources of data (see Phillips and Khachatourians 2001, for details) to use as a base for testing market adoption rates for new varieties. Downstream of farm gate, international trade flows of canola germplasm, seed, oil, and meal can be tracked through the FAO trade database, Industry Canada's Strategis export data and through the Canadian Seed Trade Association's estimates of exports of germplasm.

The Saskatoon-centred Innovation Entrepot

The dimensions of the Saskatoon innovation entrepot can best be analyzed by looking at its relative role in creating knowledge, using knowledge, and commercializing new products. While one might conclude that Canada is the main canola innovator, considering its record as the lead innovator and early adopter of all the new traits over the past 40 years, in fact, a significant share of the applied research used to develop the processes for developing new varieties was done in other

Figure 2: The Saskatoon biotechnology entrepot and its global connections

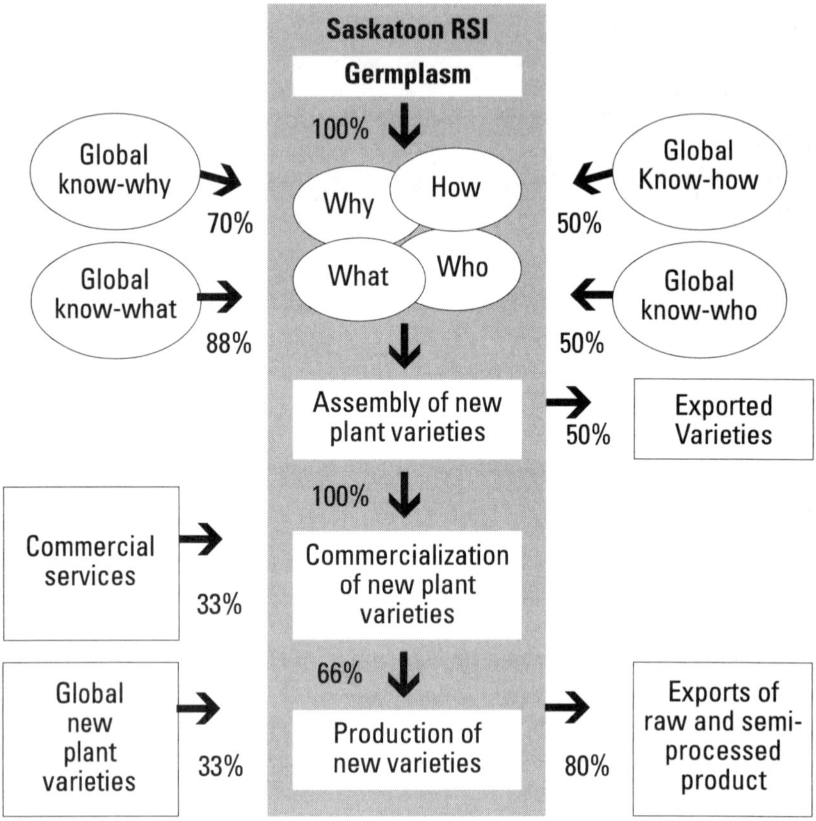

countries. Much of the applications-based research (e.g., uses for new oils) also happens elsewhere. This information suggests that Canada may instead operate in a niche in this global knowledge-based industry, as an entrepot undertaking and assembling the know-why, know-how and know-who of varietal breeding and primary production. However, the bulk of the activities upstream and downstream of that stage in the production system are now and may continue to be done elsewhere. Figure 2 illustrates the relationships between the global industry and the Saskatoon entrepot.

To illustrate this phenomenon, the Saskatoon-centred innovation system is examined in the context of knowledge creation, which involves the four "knows" and the research community, and knowledge commercialization, which involves related and supporting industries, regulators, and producers.

Table 3: Sources and uses of scientific research related to canola, by location

	Basic scientific research	Canola related agriculture research
US	36%	12%
Canada	13%	30%
UK	13%	14%
Germany	6%	9%
France	5%	7%
Japan	5%	3%
Australia	4%	3%
Sweden	3%	4%
India	3%	6%
Netherlands	3%	1%
Other	9%	11%

Source: ISI special tabulations.

Knowledge-Creation Systems

At the know-why level, Canada is the single largest country producing pure agricultural research into canola, but the basic research that underlies that effort is disproportionately coming from the U.S. (35%) and other countries. Canada is the only significant canola researching country that imports significant amounts of basic research relative to its own canola research (Table 3).

Moving forward in the research stream, Canada, particularly the public labs in Saskatoon, held a dominant position in canola-specific scientific research until the early 1980s. The combined Canadian locations contributed between 33% and 40% of the pure agricultural research on canola in the world to that time. More recently, however, the global research volume has expanded greatly while Canadian research efforts have been relatively steady. The result is that canola research has disseminated into a wide variety of locations and institutions. In 1981 only 22 countries were doing any research into canola. By the mid-1990s, more than 40 countries were doing research into canola on a continuing basis, and more than 70 countries had done some work on canola over the intervening period. By the mid-1990s, Canada's share of this research dropped to about 20%. Furthermore, as the basic scientific research has become more specialized, collaborations are playing a greater part in the global evolution of research capacity. The top nine canola research countries undertook between 5% and 22% of their

Table 4: Private and Public Owners of Patents by country of residence

	Technology Public	Private	Product Public	Private	Total patents
Total Patents	66	220	37	311	634
Residence of inventor:					
- Canada	32	18	13	12	75
- Other	34	202	24	299	559

Source: Search of CIPO Canadian Patent Database, December 1999.

research in collaboration with others in other countries (Phillips and Khachatourians 2001). Canadian researchers undertook only 8% of their effort via international collaborations (the lowest of all countries except India), suggesting that by this time Canada was more self-reliant than many other countries at this stage.

Nevertheless, Canada is far from self-sufficient. Canada is significantly more dependent on foreign actors for the proprietary know-what technologies used in the industry than for basic R&D. The Canadian patent database shows that since 1985 there have been 634 patents issued for canola-related innovations (45% for process innovations), the vast majority of them to foreign (mostly U.S.-based) research teams and companies (Table 4). Canadian researchers accounted for only 75 patents, more than half of them granted to a few federal research scientists from the AAFC and the NRC. In practical terms, virtually all of the key transformation technologies used by the canola sector are the proprietary products of non-resident companies (see Phillips 2000, for a detailed listing of owners of key proprietary know-what technologies).

Moving downstream in the product development system, field trial data provides insights into who is undertaking research, where it is taking place, and with whom. International field trial data shows that Canada had an early and dominant lead over any competing country. Although 12 countries had field tested at least one transgenic variety of canola between 1988 and 1995, Canada accounted for 54% of all field trials and performed almost four times as many trials as France, the next highest country.

The generation and transmission of the non-codified knowledge in the regional system is the key factor holding things together. People develop skills and working relationships that, together, convert bits of information into operable knowledge. This tacit type of knowledge is learned almost exclusively through experience. Researchers learn how to do things and whom to work with through trial and error. Most of the innovation literature assumes that this know-how and, perhaps more importantly, this know-who, evolves within corporations or institutions. That may hold true in an industry or within firms that are largely self-sufficient, but, as noted, there are few firms that have the internal capacity to undertake all the research and development necessary to create a marketable canola variety. Some companies may have such capacity within their global operations, but in many cases working through the multiple, geographically-dispersed layers of these multinational enterprises is more complex and less cost-effective than buying-in from a more accessible and timely local source. Hence, although Monsanto and AgrEvo (now Aventis), for instance, both have giant research "universities" and labs at their headquarters, both have collaborated extensively in Saskatoon with both the AAFC and the NRC. Furthermore, in knowledge-based industries, training and upgrading are critical, making it essential for private researchers to interact with the broader research community. For these reasons, most firms in the industry have developed an extensive community of networks with both collaborators and competitors, including other private companies, universities, AAFC and the NRC.

A 1998 survey of canola firms globally and in Canada (Phillips and Khachatourians, 2001) highlights the importance of the proximity of competitors and/or collaborators as factors in locating their research efforts. Half of all the respondents, representing the majority of private companies responding, acknowledged the importance of proximity to either collaborators or competitors. About 40% recognized the importance of being close to their collaborators, particularly the NRC and AAFC in Canada and key research universities in the U.S. (Table 5).

As with most communities, proximity matters. Formal and informal face-to-face meetings and working side-by-side on laboratory benches and in the greenhouses are critical in both developing the know-who and transmitting the know-how. It is highly unlikely that the community would have developed if there had been only competitive firms in Saskatoon; the non-competitive environment offered by AAFC and NRC has created the platform for the necessary relationships. Both AAFC and NRC have extensive arrangements with each other, with public universities, and with private companies. In 1995-96, the NRC had more than 31 arrangements, ranging from research agreements to collaborative work agreements and licences, that brought more than 65 guest researchers from other institutions into the NRC labs (NRC 1997). In 1997-98, the NRC welcomed 109 guest researchers. The key feature of

Table 5: How important are the following to your decisions to both undertake research and to locate the research in labs in Saskatoon or elsewhere?

	N = 28	%
Proximity to competitors or collaborators	14	50%
- Proximity to collaborators	11	39%
- Proximity to competitors	8	29%
Access to local pool of skilled labour	7	25%
Access to large and accepting farm market for seeds being produced	6	21%
Key scientists either in your company or in partner organizations	5	18%
Role of government agencies (federal, provincial, regional) related to hospitality, red tape (or lack of)	5	18%
Access to labs, greenhouses and test fields	4	14%

Source: Phillips and Khachatourians 2001.

these arrangements is that the core research team at NRC is able to learn from all of the collaborations, adding further to the know-how knowledge and providing a visible, efficient point of entry for know-who knowledge. Even firms not resident in Saskatoon have developed extensive links to gain access to the knowledge of those two institutions. This suggests that spill-over benefits from the know-how and know-who located in Saskatoon may be significant and may not move far from Saskatoon.

A key element in pulling the research effort together is the specialized, skilled labor force. The gradual agglomeration of canola research in Saskatoon and Canada suggests that there should be some evidence of a speciality labour market evolving. More than one-quarter of the companies that responded to the canola industry survey, and the clear majority of private companies, indicated that access to a deep, local, skilled labour pool was important. A survey of canola research employees in Saskatoon received 390 responses (from 1000 surveys distributed). They represented 169 person-years of canola-related employment in 1998, compared to the 248 person-years of related employment in Saskatchewan identified in the company survey (Phillips and Khachatourians, 2001).

When the degree-granting institutions were examined, the dynamics of the industry become pronounced (Table 6). All but two of the employees with a technical diploma were trained in Canada; 82% of the employees with technical diplomas received their training from the

Table 6: Distribution of employees in the Saskatoon agri-food research community, by degree, 1998

% total	Technical degrees	Undergraduate diploma	Masters' degrees	Ph.D. degrees
Saskatchewan	82%	66%	43%	27%
Rest of Canada	16%	25%	41%	37%
US	1%	2%	6%	10%
Europe	0%	4%	7%	18%
Other	1%	3%	3%	8%

Source: Phillips and Khachatourians 2001; survey of canola industry employee undertaken in July 1998

Saskatchewan Institute of Applied Science and Technology, mostly through the two-year Biotechnology Technologist Program in Saskatoon. The same trend is seen at the undergraduate level. All but 22 of the 85 respondents with a bachelor's degree were trained in Canada. There has been a bit more intra-country mobility at this level of training, with only 66% of the respondents acquiring their training in Saskatchewan. The Saskatoon labour market draws extensively from Ontario, Alberta, and Manitoba. At the graduate degree level, the labour market becomes significantly more mobile. Fewer than half of the employees with master's degrees and only about one-quarter of the employees with doctorates were trained in Saskatchewan. In short, the higher the education degree, the greater the mobility and cross-national movement of employees. At the Ph.D. level, more than 35% of the workers were trained offshore, in Europe, the U.S., and other countries. Nevertheless, it is important to note that the single largest source of both masters and doctoral level employees is the local university in Saskatoon.

The data shows that although the local labour market is able to supply much of the labour required, some higher-skilled employees must be recruited from elsewhere. All employees were asked what features of the job and community affected their willingness to move to or from Saskatoon. Given conventional wisdom in the industry that after-tax income and quality of life were most important, it was somewhat surprising that the thickness of the labour market was the key consideration mentioned by respondents. All employees with graduate degrees who responded to this question ranked proximity to other companies or agencies that could hire them among their top five considerations, with 87% of the respondents citing it as the most important consideration. The second most important consideration was the type of work in the job, another feature of a thick labour market. Salary and benefits

Table 7: The location of research stars by country, 1981-96

	Stars		Emerging/borderline stars	
	all scientists with 20 articles or more	at least 20 articles; >5.0 cite rate	all scientists with 15-19 articles	Scientists with 15-19 articles; >5.0 cite rate
Canada	13	5	25	13
- Saskatoon	3	2	8	4
Australia	0	0	1	1
Europe	13	8	9	6
- France	3	1	1	1
- Germany	4	2	2	0
- Poland	1	0	0	0
- Sweden	2	2	1	1
- UK	3	3	5	4
Japan	3	2	1	1
US	0	0	4	4
Total	**29**	**15**	**40**	**25**

Source: ISI special tabulations

came third, followed closely by career prospects. Almost all the other factors, either related to the job or related to the community, were ranked well below these four factors (Phillips and Khachatourians 2001).

Zucker, Darby, and Brewer (1998) offer an approach to determining more explicitly how the labour market contributes to agglomeration. Their study examined the role of human capital in the birth of U.S. biotechnology enterprises by looking for causalities between the location of research stars and the creation of new firms. They defined "stars" as scientists who had discovered 40 or more genetic sequences or written 20 or more articles on genetic sequence discoveries. They concluded that the presence of active stars in a region was strongly positively correlated with the start-up of new ventures, stating that "at least for this high-tech industry, the growth and location of intellectual human capital was the principal determinant of the growth and location of the industry itself" (Zucker, Darby, and Brewer 1998, 302).

If we take "stars" to be those who publish at least 20 articles on canola and "borderline stars" as those who publish 15-19 articles, we find 69 individual scientists worldwide who fit the criteria (Table 7). About 45% of the stars are in Canada, 45% in Europe and 10% in Japan. Approximately 63% of the borderline stars are in Canada, with the rest scattered in Europe, Japan, the U.S., and Australia. In total, the 69 stars and borderline stars, just less than 1% of all the scientists working on canola, produced 1,523 articles, or about 31% of all the articles produced over the period (Zucker, Darby, and Brewer, 1998, found that the stars in their study represented 0.75% of all scientists but 17% of all articles). The largest single geographic concentration of stars and borderline stars in the world is in Saskatoon, where 11 (16%) of the scientists live and work. If the stars and borderline stars are then assessed by their citation rates, Saskatoon has 6 of 40 (15%) and about one-third of all the Canadian stars and borderline stars.

One could conclude from this analysis that during the knowledge creation phase, Saskatoon and Canada are significantly dependent on global markets for know-why and know-what, but that they are increasingly self-sufficient as research becomes more applied.

Knowledge Commercialization Systems

Economic theory indicates that firms that locate where there are extensive backward linkages into supporting industries and forward linkages into the market can realize economies of scope. In Saskatoon, however, apart from publicly provided infrastructure, there is limited evidence of any existing or developing critical mass of other specialized industries that strengthen the bond between the canola research industry and the location.

A number of public investments in Saskatoon support the industry. These include Innovation Place research park, the Saskatchewan Research Council facilities, and the POS Pilot Plant, which undertakes scale-up work on oil and meal properties for companies with new varieties of canola and consults on oil processing technologies. Support also extends to financing. Available private venture financing is limited, with no access to a local stock market (Calgary is the nearest market) and with only limited venture capital lending. The public sector fills the gap. Two financial institutions, the Royal Bank in Saskatoon and the CIBC in Winnipeg, both in partnership with Western Economic Diversification, have specific knowledge-based lending facilities. While there is limited public lending capacity through Western Economic Diversification, the Business Development Bank of Canada, the Saskatchewan Opportunities Corporation and the Agri-Food Equity Fund all have offices in Innovation Place or in Saskatoon. Taken together, the operation of the various federal, provincial, regional, and local tax and fiscal programs has sharply reduced the cost of locally conducted research and development, to the point where, in effect, much of the private innovation enters the market tax-free (Phillips and Khachatourians 2001). Publicly-provided infrastructure, company-specific industrial incentives, federal and provincial matching research funds, research in the public labs, public financing of university research, and a plethora of grants and tax incentives combine to underwrite virtually all of the out-of-pocket costs for private research in Canada. Furthermore, the privileged tax status of western Canadian farmers ensures that producers retain most of the benefits they gain from early adoption. Hence, in spite of weak private capital markets, the public system has effectively made the biotechnology sector more attractive through granting it nearly tax-free status.

Although the financial, accounting, and legal communities have recently been restructuring to service the growing private research effort in Saskatoon, surveys conducted in 1998 suggest that only limited accommodation had been made by then to support and service the developing biotechnology industry (Phillips and Khachatourians 2001). All of the multinational firms undertaking canola research in Saskatoon indicated in discussions that they did their banking as part of their corporate effort, with much of the service coming from their Canadian or global headquarters, which were always located in another city or province. Meanwhile, the Chartered Accountants of Saskatchewan, the Certified Management Accountants of Saskatchewan, and the Certified General Accountants of Saskatchewan indicated that no specific listings are available for practices that deal specifically in the area of biotechnology. A fax-back survey of biotechnology companies undertaken in August 1998 revealed once again that multinational firms used accounting services purchased through their Canadian head offices, while the smaller firms used local accountants and auditors (Phillips and Khachatourians 2001). None of the accounting practices were identified as having a speciality in accounting for knowledge-based enterprises.

A third key service required by knowledge-based firms is legal support for protecting intellectual property that may have commercial value. A key limiting factor in Saskatoon is that there are no resident patent agents in Saskatchewan. Most of the patent agents operating in Calgary, Edmonton, Winnipeg, or Ottawa have relationships with local legal firms or have liaison offices in Saskatoon to link to the local demand for the service. Few of the legal practices in Saskatoon have the capacity to support firms in protecting their intellectual property. While 9 of the more than 50 multi-partner legal practices operating in Saskatoon indicated they had some capacity in the area of intellectual property rights, only two firms indicated they had an ongoing practice.

In addition, the focus of the local legal community is on Canadian law, which limits their usefulness, as innovators wish to protect most of their innovations internationally. In short, firms are being serviced, but the presence of many multinationals, which buy these services internationally, combined with the absence of any registered patent agents in the province and the small number of full-time lawyers specializing in intellectual property law limits the scope of benefits that the industry could accrue.

Moving downstream from the research stage, there are vital forward linkages that increase Saskatoon's attractiveness as a research site. More than one-fifth of all enterprises surveyed (Table 5), representing most of the larger breeding operations, indicated that rapid access to a receptive seeds market was critical to their decision to locate in Canada. They confirmed that undertaking research and commercializing the resulting canola varieties under the same regulatory system was a key feature in their location decisions. Heller (1995) estimates that a regulatory delay of one year decreases the rate of return for a biotechnology product by 2.8%, while a two-year delay decreases the rate of return by 5.2%.

Both the regulatory system and market responsiveness do matter (Porter 1990). In the first instance, the regulatory system determines how quickly farmers can and will adopt new varieties. The Canadian regulatory system is generally viewed as somewhat slower than the U.S. system (although this lag may have diminished with the experience gained with early transgenic varieties), but it is clearly faster than in either Europe or Australia, where no transgenic canola varieties were approved and produced as of December 2001. The first transgenic varieties were approved for unconfined release in 1994 in both the U.S. and Canada and began commercial production in 1995. Farm programs also contribute to the pace of adoption of new varieties. In Canada, new varieties of canola are automatically eligible for crop insurance and other stabilization assistance, without limits on acreage seeded or location. In the European Union (EU), where the Blair House Accord and the World Trade Organization (WTO) agreement limit canola acreage, incremental planting of new canola varieties is treated less favourably than existing acreage, which limits adoption in some countries and regions. Furthermore, canola is not yet suited agronomically to most parts of the U.S. and Australia, which limits adoption there.

Farmers base their response to canola varieties partly on the regulations, partly on the markets, and partly on their prior knowledge and experience in adopting new varieties. There is compelling evidence that farmers have substantial experience that encourages adoption. Farmers have rapidly adopted the new varieties, with an estimated 77% of acreage in 2000 planted in herbicide-tolerant varieties.

Extensive cooperative farm service networks have helped to create this large, receptive and relatively sophisticated farm market for new seeds in the prairies. The Prairie Pools, in particular, with a historical delivery share of about 60% and membership that includes the majority of farmers in the West, have aggressively positioned their organizations as wholesalers for new varieties. They partnered extensively with Svalof in earlier years and more recently with AgrEvo and Monsanto to deliver their proprietary herbicide-tolerant seeds to their members. In addition, the extensive network of not-for-profit producer and industry organizations, such as the Canola Council of Canada and the various provincial growers' associations, ensure the rapid and efficient adoption of the technology.

Beyond the wholesale system, however, the market opens up to the rest of the world, with little evidence that Saskatoon, or for that matter Canada, dominates the system. The 20% of the product consumed within Canada is marketed through the Winnipeg Commodity Exchange or through a number of proprietary processing and retailing supply chains, with few backward links to the R&D community in Saskatoon. Offshore, Canada's influence is limited. Foreign-owned companies produce all of the resulting end products, using their own proprietary technologies.

The net result appears to be that while linkages in the knowledge-creation system are the base for this innovation cluster, the downstream capacity to commercialize the product is quite extensive and is one of the key factors contributing to the location and expansion of this activity in Saskatoon and western Canada.

Conclusions and Implications

This paper has presented a modified framework for examining modern innovation. It has also provided some compelling, if not conclusive, data from the Saskatoon-centred biotechnology-based innovation cluster which supports the view that innovation systems should be more properly examined as trade entrepot rather than self-contained, self-sustaining centres. In Saskatoon's case, the innovation cluster has found a niche as the developer and early adopter of new transgenic canola varieties. The university, cooperative and government research labs and sophisticated producer and marketing organizations are at the heart of the cluster. However, it is clear from the evidence that the innovation cluster is not in any way independent or self-sufficient. Rather, it draws heavily on global R&D, worldwide labour markets and speciality services from across North America. Thus, less than half of the value added

Table 8: Global canola production and seed trade, 1961 and 1998

	share of global production		export market share	
	1961	1998	1961	1998
Australia	0%	4%	0%	8%
Canada	7%	21%	40%	47%
China	11%	23%	0%	0%
European Union (15)	15%	27%	11%	37%
– France	3%	10%	10%	28%
– Germany	7%	9%	1%	4%
United Kingdom	0%	4%	0%	3%
India	37%	14%	0%	0%
United States of America	0%	2%	0%	3%
World totals (thousands metric tonnes)	**3,596**	**35,869**	**310**	**8,717**

Source: FAO 1999.

to the product is added locally. In short, it operates in remarkably similar fashion to a traditional trade entrepot.

The returns for success are significant. In the canola case, as the research effort increased, rapeseed/canola production shifted toward countries that intensively managed canola as a knowledge-product and away from countries that did not compete on the knowledge front (excluding China, which has not allowed market forces to determine production) (Table 8). Together, the EU and Canada, both leaders in the research and development of canola, have more than doubled their share of global production to 45% over the past 30 years. Meanwhile, India, Pakistan, Poland, and Japan invested little in R&D for new canola varieties and saw their share of global production drop from 50% in 1961-65 to about 21% in 1992-99. In addition, a number of new producers are on the horizon with commercial quantities of canola. The U.K., the U.S., and Australia, each significant investors in canola-related research, entered and significantly expanded their market shares during the 1980s (amounting to 8% of global production in 1992-99). Although none of these countries is producing enough yet to challenge the EU or Canada, their future roles cannot be ignored, especially at the product end of the business. Canada sustained and solidified its commanding position in the seed export market throughout the period. As Canada's share of production rose, so did its share of trade, reaching 47% in 1998. Europe, the only other large exporter, saw its share of trade rise from 11% to 26% over the same period. Meanwhile, as research and production was relatively stagnant in most of the world's other canola producing countries, their export market share dropped, to 14% in 1996 from 49% in 1961.

There are three aspects of this model that warrant further discussion. First, the concept of knowledge-based clusters operating as entrepot needs to be examined in other areas to determine whether the Saskatoon biotechnology cluster is a general or specific case. Other regional clusters should be examined using this (or a modified) methodology to determine whether they are self-sufficient or exhibit entrepot characteristics. A good example of a related study is the work of Feldman (1994), which examines the role of Johns Hopkins University in the Baltimore area. Other investigations need to be undertaken.

Second, the model needs further refinement. The definitions and methods for evaluating the knowledge factors could be improved. Furthermore, while serendipity, the wild card in the mix, may not be amenable to modelling in the strictest sense, it may be further investigated in the context of examining the know-who dimension. There is evidence in the literature on creativity that communication, connections, and critical mass play a significant role in the occurrence of serendipitous outcomes, and that while such outcomes cannot be planned, the conditions for increasing their likelihood can be nurtured.

Third, this alternate model of innovation poses some serious challenges for development policy. Much of the current development effort has a strong mercantilist orientation, with a focus on generating exports while impeding imports. All levels of government in many countries are actively using their tax and fiscal policy to encourage greater local R&D or to attract global firms to relocate their R&D programs into their jurisdiction in an effort to generate higher value exports or to replace imports. This effort has often involved preferential support for national champions or exclusive deals to encourage MNEs to relocate their activities. Usually governments do this without any consideration of the corresponding relationships and interactions that firms, especially knowledge-based ones, require to succeed. If innovation can be thought of as being confined within a firm or a regional or national community, then such a narrow approach might have some chance of succeeding. But if innovation is truly global, as appears to be the case in many of the life sciences, narrow mechanistic self-sufficiency strategies may fail or prove to be counterproductive. The evidence from the biotechnology sector suggests that innovation is truly global, which goes a long way toward explaining why both firms and skilled employees are more interested in the innovation community than in fiscal incentives, public infrastructure, or government supports. By extension, a

Table 9: Public policy options for nurturing knowledge-based innovation clusters

Knowledge factor	Policy prescription
Know-why	• Develop absorptive capacity through basic research capacity in universities and public labs • Nurture two-way international flows of information through programs that support and encourage international collaborations
Know-what	• Create effective intellectual property protection systems that facilitate two-way international flow of innovations. • Encourage location of competing and collaborating multinational enterprises to encourage transfer of proprietary technologies
Know-how	• Develop and maintain a critical mass of researchers and technologists focused on advancing and using the technologies either in public labs (e.g., AAFC) or in private firms • Nurture "thick" labour markets through post-doctoral research support and liberal labour policies
Know-who	• Develop and maintain open-platform institutions to facilitate research collaborations (e.g. NRC) • Nurture collaborative, industry-led networks to facilitate communications and to assist with developing the forward and backward institutions necessary for efficient commercialization

mercantilist policy that discourages global links not only might fail to attract but could ultimately drive out firms or researchers as they seek access to the global community.

The entrepot innovation model, as illustrated by the Saskatoon-centred, biotechnology-based innovation cluster, offers alternative policy options (Table 9). One key to succeeding in this type of world would be to invest in those institutions and mechanisms that encourage the development of and access to the four knowledge factors which provide the true base for the absorptive capacity of a research economy. From this example, it is possible to identify a number of elements vital to creating that capacity. First, there must be effective mechanisms to practically and legally transfer knowledge, which at a minimum involves a domestic research community with international collaborations, some

MNEs with proprietary technologies, and appropriate legal protection for intellectual property. Second, there must be open and accessible labour markets for skilled workers. Third, there must be a platform, such as the NRC, AAFC, and industry-led networks, in which community-based interaction and synergies can develop. These elements provide the foundation for absorbing global knowledge. Others, such as preferential financing and speciality commercial services, may be important but appear to be second-order requirements.

In short, innovation clusters are very attractive economic development tools, but they must be nurtured with an appreciation for their partial and incomplete nature. Fundamentally, they are part of a global innovation system, and cannot thrive if cut off from the lifeblood of the system – ideas, skilled labor, and collaborative platforms.

References

Andersen, E. and Broendgaard, A. (1992). Integration, innovation and evolution. In *National Systems of Innovation: Towards a Theory of Innovation and Interactive Learning,* ed. B. Lundvall, 242-64. New York: Pinter.

Arrow, K. (1962). Economic welfare and the allocation of resources for invention. In *Rate and Direction of Inventive Activity*, 609-26. Princeton: NBER and Princeton University Press.

Bannock, G., Baxter, R., and Rees, R. (1972). *The Penguin Dictionary of Economics*. London: Penguin Books.

Canadian Food Inspection Agency (CFIA). Canadian Varieties, January 1, 1923 to June 24, 1998. *Special tabulation from the Plant Health and Production Division.*

Canadian Food Inspection Agency (1999). Summary of Experimental Releases. *Special tabulation for Brassica napus and rapa.*

Canadian Intellectual Property Office (2000: November). *Patent Guide.* (Web page) http://strategis.ic.gc.ca/sc_mrksv/cipo/patents/pt_main-e.html

Chesnais, F. (1992). National systems of innovation, foreign direct investment and operations of multinational enterprises. In *National Systems of Innovation: Towards a Theory of Innovation and Interactive Learning,* ed. B. Lundvall, 265-95. New York: Pinter.

Dalum, B. (1992). Export specialisation, structural competitiveness and national systems of innovation. In *National Systems of*

Innovation: Towards a Theory of Innovation and Interactive Learning, ed. B. Lundvall, 191-225. New York: Pinter.

Endacott, G. (1964). An Eastern Entrepot: A collection f documents illustrating the history of Hong Kong. *Overseas Research Publication 4*: 293. London: HM Stationery Office.

Fagerberg, J. (1992). The home market hypothesis re-examined: the impact of domestic user-producer interaction on export. In *National Systems of Innovation: Towards a Theory of Innovation and Interactive Learning*, ed. B. Lundvall, 226-41. New York: Pinter.

Feldman, M. (1994). The University and Economic Development: The Case of Johns Hopkins University and Baltimore. *Economic Development Quarterly* 8(1): 67-76.

Food and Agriculture Organisation (FAO) (1999: November). FAOSTAT Data. (Web page) http://www.fao.org/.

Freeman, C. (1987). *Technology and Economic Performance: Lessons from Japan*. London: Pinter Publishers.

Gray, R., Malla, S., and Phillips, P.W.B. (1999). *The Effectiveness of the Research Funding in the Canola Industry*. Saskatchewan: Saskatchewan Agriculture and Food.

Grossman, G. and Helpman, E. (1991). *Innovation and Growth in the Global Economy*. London: MIT Press.

Heller, J. (1995). Cited in Goudey, J., and Nath, D. (1997). *Canadian Biotech '97: Coming of Age*. Toronto: Ernst & Young.

Industry Commission (1995). *Research and Development*, 3 Volumes. Canberra: Australian Government Publishing Service.

Institute for Scientific Investigation (ISI) (1997: November). Citations database, special tabulation of academic publications based on key word search for canola.

Klein, S. and Rosenberg, N. (1986). An overview of innovation. In *The Positive Sum Strategy: Harnessing Technology for Economic Growth*, ed. R. Landau and N. Rosenberg. Washington: National Academy Press.

Krugman, P. (1998). What's new about the new economic geography? *Oxford review of Economic Policy* 14(2): 7-17.

Lundvall. B., ed. (1992). *National Systems of Innovation: Towards a Theory of Innovation and Interactive Learning*. New York: Pinter.

Malecki, E. (1997). *Technology and Economic Development: The Dynamics of Local, Regional and National Competitiveness.* Toronto: Longman.

Marshall, A. (1890). *Principles of economics.* London: Macmillan.

McLeod, A., ed. (1974). *The story of rapeseed in Western Canada.* Regina: Saskatchewan Wheat Pool.

Moed, H., Burger, W., Frankfort, J., and van Raan, A. (1985). The use of bibliometrics data for the measurement of university research performance. *Research Policy* 23(2): 187-222.

National Research Council (1992). *From Rapeseed to Canola: The Billion Dollar Success Story.* Saskatoon: National Research Council.

National Research Council (1997). Bi-Annual Report: 1995-97. (Web page) http://www.pbi.nrc.ca/96annrpt/bus.html.

Nelson, R. (1988). Institutions supporting technical change in the United States. In *Technical change and economic theory*, ed. G.C. Dosi, C. Freeman, R. Nelson, G. Silverberg, and L. Soete. London: Pinter Publishers.

OECD (1996). *The Knowledge Based Economy.* Paris: OECD. (Web page) http://www.oecd.org/dsti/sti/s_t/inte/prod/kbe.htm.

Phillips, P. W. B. (2000). Intellectual property rights and public research in Canada. In *Agriculture and Intellectual Property Rights: Economic Institutional and Implementation Issues in Biotechnology*, ed. V. Santaniello, R. Evenson, D. Zilberman, and G. Carlson. Wallingford, U.K.: CAB International.

Phillips, P.W.B. and Khachatourians G.G., ed. (2001). *Biotechnology's impact on Global Agriculture and Food production: Innovation, invention and investment in the canola sector.* London: CAB International.

Porter, M. (1990). *The Comparative Advantage of Nations.* New York: Free Press.

Romer, P. (1990). Endogenous technological change. *Journal of Political Economy* 98(5:2): S71-S102.

Schumpeter, J. (1954). *Capitalism, socialism, and democracy.* London : George Allen & Unwin.

Solow, R. (1956). A contribution to the theory of economic growth. *Quarterly Journal of Economics* 70(1): 65-94.

Trajtenberg, M. (1990). *Economic Analysis of Product Innovation: The case of CT scanners.* Cambridge: Harvard University Press.

Zucker, L., Darby, M., and Brewer, M. (1998). Intellectual human capital and the birth of U.S. biotechnology enterprises. *American Economic Review* 88(1): 290-306.

Chapter 3

Techmaps: A Tool for Understanding Social Capital for Technological Innovation at a Regional Level

Richard Smith

Abstract

This chapter looks at recent examples of the use of a variety of techniques to describe relationships and the transfer of technology and people between and among high technology firms in Canada. In each case the focus of the resulting map or tree is at the regional or sub-regional level. The chapter describes two examples of this mapping process and then considers these efforts in light of recent literature on social capital and regional systems of innovation. The chapter concludes with suggestions as to how future Techmaps might be adapted to provide insights from those areas.

Introduction

In Canada, as in jurisdictions around the world, there is considerable debate and discussion on the topic of innovation. Government agencies, in particular, have been asked to create a climate hospitable to innovation and otherwise enhance innovativeness in firms. However, to create and manage policy measures that encourage or foster innovation, it is necessary to know who is involved and where the innovative organizations are located. One proposed response to this problem is a genealogy or map of the high technology firms in a region or sub-region. In Canada, one of the earliest of these was completed in 1995 and takes the form of a family tree of computer and communication firms in the Ottawa Valley.[1] That map was created by consultant Denzil Doyle, and has been widely distributed in Canada.

---

[1] According to Adam Holbrook, there was a precedent for the Doyle map:
"As I remember it the first geneology was drawn in around 1980 by Roger

Doyle's map was demonstrated at a Vancouver, Canada conference in December 1996. Soon afterward, representatives of a high technology consulting firm (PricewaterhouseCoopers), a regional coordinator of networking, financing and mentoring programs (B.C. Advanced Systems Institute, or "ASI"), and a university research centre (the Centre for Policy Research on Science and Technology at Simon Fraser University, also known as CPROST) got together and agreed to create a similar project for their region. The result is the Techmap. It was researched, designed, and produced in the spring, summer, and fall of 1997. A follow-up project, known as Techmap II, was developed in the fall of 1998.

The Techmap projects generated considerable interest among policy makers and the high technology community in British Columbia. PricewaterhouseCoopers went on to create similar maps in several other Canadian regions, including Ottawa, Kitchener-Waterloo, Montreal, and Toronto. The creation of these maps was seen as a way to raise the profile of the company as a service provider in those communities, and to contribute to greater awareness of the high technology industry in Canada.

From the perspective of B.C. Advanced Systems Institute, the map provided a novel way to illustrate the importance of one of their main foci, the networks among high technology firms and between technology firms and universities.

The project was also of interest to researchers at the Centre for Policy Research on Science and Technology, because of our interest in regional innovation systems. Regional systems of innovation is an area of study that has emerged out of the work on national innovation systems by Lundvall and others (Lundvall, 1992). In Canada, as in other regionally diverse jurisdictions around the world, the study of innovation systems has been expanded to include regional innovation systems (Acs, de la Mothe, and Paquet, 1996). One example of these initiatives, Innocom, involves researchers at several universities in western Canada. It is part of the Innovation Systems Regional Network (ISRN), sponsored by

---

Voyer of MOSST who at the time was trying to prepare a defence for the minister over the failure of Microsystems International Ltd (see the Ottawa Techmap). This was a semiconductor / integrated circuit firm into which [the federal government] put $300M but which went bankrupt in the face of superior technology in from Asia. What [Voyer] showed was how MIL led to the companies such as Mitel and the precursors of Newbridge (although Newbridge did not exist then). It was done strictly for policy/political reasons, not for any research capability. Voyer's chart (a small 8 1/2 by 11 sheet) was never published since it was part of a briefing note. You are correct in saying that Doyle's map was the first one published." Adam Holbrook, correspondence, 24 May 2000.

National Research Council, the Social Sciences and Humanities Research Council of Canada, and the Natural Sciences and Engineering Research Council of Canada.

The premise of both regional and the national studies of innovation systems is that innovation – the process of bringing new ideas to market – is a complex interaction of a number of players, including universities, government labs, and private companies. For any researcher interested in the innovation that is fostered by collaboration among firms (Smith, Takanashi, and Ahmed 1998), the map provides a good visual representation of the widespread nature of collaboration and the results of that collaboration. In addition, these maps now appear to be a way to provide insight into the role of social capital in innovation.

Social Capital

The study of social capital has its roots in sociological studies of community, especially community economic development (Bourdieu 1986). As Jane Fountain and others have begun to report, there is a useful link between social capital and innovation in science and technology (Fountain 1998). Social capital can be defined as "...the sum of the actual and potential resources embedded with, available through, and derived from the network of relationships possessed by an individual or social unit" (Nahapiet and Shoshal 1998, 243).

As others have pointed out, social capital is not a unidimensional concept, and it is useful to think of at least three major aspects to social capital. These include the structural, relational, and cognitive dimensions (Nahapiet and Shoshal 1998). Nahapiet and Shoshal point out that, following Granovetter (1973), the structural aspects of social capital concern "the properties of the social system and the network of relations as a whole" (1998, 244).

> The term [structural dimension] describes the impersonal configuration of linkages between people or units. ...the structural dimension of social capital refer[s] to the overall pattern of connections between actors – that is, who you reach and how you reach them. Among the most important facets of this dimension are the presence or absence of network ties between actors; network configuration or morphology describing the pattern of linkages in terms of such measures as density, connectivity, and hierarchy; and appropriable organization – that is, the existence of networks created for one purpose that may be used for another (Nahapiet and Shoshal 1998, 244).

The role of the maps in understanding social capital was not a motivator for the original research. Once we looked at the map and the way in which it was used we realized that people were seeing in it something beyond a description of a system of innovation. In fact, it was the popularity of the maps that began to make us think that they might be useful for illustrating social capital in a region. People involved in high technology industries were very interested in having a copy of their map, and locating themselves on it, showing their connections to other players.

The effort to increase the richness of the maps is described in more detail in a later section, but first the story of the first Techmap must be told.

The First Techmap

Once we had decided to create a Techmap, the next step was to figure out how to do so. To some degree, describing the relationships between firms is like creating a genealogy. In the case of a real family, there are documentary records (e.g., birth certificates, baptismal records, family bibles, and gravestones) and oral history. Tracing the genealogy of the high technology "family" in British Columbia relied on a great deal of oral history in the form of extended interviews with individuals who had been active in the community for a long time. The first of these interviews, in March 1997, resulted in information on nearly a dozen firms in the telecommunications, computing, and telephone industries. The information was transcribed onto large sheets of paper that provided a graphic representation of each firm, its relatives (spin-off companies), technologies, and employees. Although these drawings were similar to a family tree, they were not a complete analog, in part because there are not always (just) two "parents" to a new firm.

This initial interview was the basis for a planning and evaluation meeting of the steering committee for the research. Strategies were worked out to contact key firms and organizations that were thought to be at the roots of the local family tree of the British Columbia regional innovation system. An interview protocol was designed and tested.

Once an interview protocol was established and pilot tested on a few individuals, graduate student researchers were trained and sent into the field to gather more information. Their method of locating respondents relied on a snowball approach (asking informants for information on other individuals who could be important sources of information) as well as ongoing suggestions from a board of advisors. We moved through the high technology sector in stages, gradually moving from segment to segment. For example, we spent a lot of time working on telecommunications equipment before moving on to the biotechnology and software sectors. This ensured that the researchers could become aware of the peculiarities and specifics of each industry.

At any given time in British Columbia the bulk of the firms in any industry are small and recently created. This is particularly true of high technology firms. In an effort to keep the task and workload manageable, we decided to limit our genealogy to firms with more than five employees that had been in business for more than five years. This decision was made with an eye on data representation, as much as on data collection, since we were cognizant of the fact that we would have to represent this information on a single sheet of paper, even if it was to be a poster-sized sheet.

Data was collected on paper forms and transferred to a simple database program. Information was collected on firm characteristics (revenues, number of employees, year incorporated, stock ticker symbol, industrial classification, etc.) as well as on the nature of a firm's relationship with other firms in the database. For this initial research, relationships were characterized as being based on transfers of technology (e.g., a license) or transfers of people (e.g., a purchase or spin-off). Information on the origins of the firm was collected from founders wherever possible, as well as from public sources and other documentary records. Relationships were also identified by date of start and end.

Once the data had been collected, a rigorous process of fact checking and verification began. As one might imagine, when the origins and relationships (spin-off history, for example) of a firm are being described, there is considerable potential for variation in interpretation of history. As many high technology companies are relatively small and many are privately held, the historical record is not always well documented. Interview sources, follow-up phone calls, faxes, and e-mail were all used to collect, craft and correct the story of each firm. Each firm was asked to sign off on the results of the research on their organization and its history.

The focus of the research was on relationships and not on firms. As a result, the database included information on each connection in our sample of over 400 firms. This resulted in a database of over 1500 relationships. Some firms had just one relationship, while others were the centre of a web of relationships including spin-off companies, transfers of technology, and transfers of people. The traditional genealogical approach, which starts at one side (or top/bottom) of a piece of paper and works to the other side, would have been overwhelmed by such complexity. Fortunately, and with the help of a design firm, we were able to come up with an alternative method for representing this information. Older firms were placed in the centre of a series of concentric circles;

Table 1: Techmap Results

In addition to the chronological progress of BC's technology industry, the BC Techmap™ fosters observations regarding the nature and direction of industry development.

Major Observations Are: *Source : BC Techmap 1997*

1. Unlike the technology industries of Ottawa, Oregon, Washington, or Silicon Valley where several large organizations spurred development, BC's technology industry is the result of many diverse influences. Some of the key influencers identified are BC Tel (via Lenkurt Electric and MPR Teltech), Macdonald Dettwiler & Associates (MDA), Glenayre Technologies, Nexus Engineering, IBM (as an employer), Simon Fraser University, University of British Columbia, and International Hydrodynamics.
2. The technology network & communications sector is strong in BC, while the life sciences and multimedia/entertainment sector appear to be the fastest growing segments.
3. The BC industry is export-focused with a high percentage of products and services finding their markets in the United States, Asia, and Europe.
4. BC technology companies continue to be acquired or be target investments as evidenced by the interest of significant Canadian, U.S., and international companies. For example, the ATM group of MPR Teltech was acquired by Northern Telecom while the WAVE group became part of IBM. Consumers Software was purchased by Microsoft and Mobile Data International (MDI) became part of Motorola. The studio of Sanctuary Woods Multimedia Corp. became part of Disney Interactive, and Distinctive Software Group was purchased by Electronic Arts. Basic Software Group (AccPac) was purchased by Computer Associates.
5. Many of the original industry influencers are no longer part of the local technology community either been bought by an outside organization or having ceased operations. This is dissimilar to the technology communities of Ottawa, Washington, and Oregon where original influencers have grown, fostered other companies, and served as anchors of the industry.
6. The size of BC technology companies today is typically smaller than those in the past. Boeing Canada employed over 10,000 people in 1943 while there were 3,400 Microtel employees at the company's peak. Lenkurt Electric had nearly 2,000 employees, Okanagan Helicopters 1500, and MDI over 1000 employees at their respective peaks. Apart from BC Tel, there are currently few large technology employers. Information Systems Management BC (ISMBC), a joint venture between BC Tel and IBM, is the largest employer with 1,300 employees. Glenayre Technologies, now a U.S. based company, has 900 U.S. employees. CREO and MDA have close to 1,000 employees each. Today the BC industry is mostly made up of small to medium sized firms.

newer firms were placed further and further out from the centre, depending on their age. Firms were grouped into regions of the overall grid, based on their networks/connections. For example, all firms related to MPR Teltech/BC Tel are in one general area to avoid a spaghetti of lines.

The first map was presented to the public in November 1997. It was widely distributed in the form of a fold-out brochure as well as a smaller number of rolled-up maps suitable for hanging on walls. Analysis of the results of the mapping process was put into the form of case studies and summaries of results, and these were printed on the back of the fold-out version of the map. The results were mainly descriptive and used some comparisons with similar research in other jurisdictions (Ottawa Valley, Washington, Oregon, Silicon Valley). Table 1 provides a sample of the Techmap results.

The focus of the first Techmap was on identifying initiators and influencers of technology-based growth in British Columbia. The universities, in particular, were found to be major initiators and influencers of the technology sector.

Companies can be related or connected to one another in a number of ways. The most obvious is by common ownership. Firms may have major shareholders in common, or they may be related by acquisition. It is common practice in the technology industry for a firm to spin-off companies from internal development projects. That was the case with the Nexus Group of companies in the late 1980s.

Another way that firms can be related is through the movement of personnel. For the first B.C. Techmap,™ we looked at the founders of a company and where they had worked before they started the new firm. For example, ex-employees of Glenayre Electronics have started a number of firms over the past 20 years.

The Techmap helps illustrate how linkages made by the transfer of people (i.e., dotted lines) are distinguished from linkages made by transfer of technology (i.e., solid lines). It also uses icons and colors to distinguish between the various types of firms (advanced devices, networks and communications, software, and so on). For example, as an originating firm, Chisolm Industries is colored orange on the Techmap™. Glenayre has an icon which graphically depicts that it is a firm with more than 500 employees.[2]

One benefit of this type of research is the ability to demonstrate that, even when a major company fails or folds, knowledge is not totally lost to the community. Thus, the section of the Techmap that depicts MPR Teltech (closed by BCTel in 1997) illustrates that much of the knowledge developed by the company remains behind in the form of knowledge held by the former employees of the company. These employees start new companies, often with the help of their former colleagues. For example, PMC-Sierra, one of the most valuable companies in British Columbia, was a spin-off from MPR Teltech.

The Second Techmap

In the summer of 1998, the original partners in the Techmap project, B.C. ASI, PricewaterhouseCoopers, and Simon Fraser University, agreed to move forward with a program of research to support the creation of "Techmap II," a new version of the Techmap. In this section we identify the objectives of that research and the results.

Objectives

The research program started in July 1998 and ended in December 1998. The program had three objectives. The first objective was to revise the Techmap methodology and to gather additional information. The second objective was to explore new options for data gathering, with a specific focus on web-based and fax-back systems for collecting data. The third objective was to build up our database of high technology firms in B.C.

First objective: New questions and refinement of the research instrument

The first objective was important because we wanted to deepen our understanding of innovation management and technology networking among B.C.'s high technology firms. We were also interested in formalizing the research methodology, which had been largely ad hoc in the first Techmap iteration.

The three research partners had different areas of interest with regard to deepening understanding. At CPROST we were interested in understanding the formation, maintenance, and exploitation of social capital in the province. As we noted earlier, Fountain has pointed out the proven link between social capital and economic performance of firms and of regional systems of innovation (Fountain 1998).

To understand better the components of social capital (norms, networks, and trust), we spent considerable time reviewing the literature

[2] Descriptions of firms are based on the original Techmap data and not on current figures. Glenayre, for example, decided to dramatically scale down its Vancouver operations in 2001.

on social capital as well as the role of knowledge management in the innovation process. We were trying to extend our understanding of the connections between firms, which in the current Techmap are represented as linkages of only two types, people or money.

For our partners, PricewaterhouseCoopers and B.C. ASI, deeper understanding also included the financial performance of technology firms and their use of government programs.

Second objective: Data collection experiments

The second objective was to improve our ability to gather Techmap information. In order to collect and manage a growing amount of information, it was important to explore data collection options beyond face-to-face interviews and single-user, flat-file databases. These techniques had proved highly effective for the first Techmap round, but we felt that the increased complexity of information being gathered and the increased number of target firms would overwhelm such an approach.

We decided to experiment with surveys that could be delivered face to face, by mail, in a fax-back format, and online. To support the data that these surveys would generate and to allow multiple simultaneous researchers, we decided to experiment with building a secure online database. The online database had to be capable of: (1) being accessed by a number of people at once; (2) being linked with the worldwide web for both data dissemination and data entry; and (3) providing good performance and security under those conditions.

Third objective: Data growth and refinement

The third objective was to expand the total number of firms in the Techmap database from approximately 400 in the original map to 500-700 firms. This database would then be used to produce the Techmap II poster. However, the poster production itself was not part of this research. The second objective was important because we sought to broaden our understanding of the technology industry in B.C. We did this by: (1) including more firms that could not be accommodated in the first map; (2) including firms that had emerged since the creation of that map; and (3) including firms in categories such as new media that hadn't been considered as part of the high tech industry during the first data collection process.

Results

Revision of methodology (Objective 1)

We expanded the scope of the project by including questions from recent assessments of best practice in knowledge management (Amidon, 1997) and innovation management (Chiesa, Coughlan, and Voss, 1996). Describing the initiators and sources of high technology firms, as we did in the first Techmap, is a necessary first step in formulating policy for technology growth and sustainability. The next step was to enhance the climate for all firms. For this reason, we sought to know more about the ways in which technology firms in B.C. managed their innovation processes and to compare this information to best practices from the literature.

The research process began with the development of survey instruments that would allow us to delve deeper into innovation management and social capital. We also sought to include questions which would help B.C. ASI better understand the extent to which B.C. high technology companies were aware of, and taking advantage of, government programs and initiatives designed to help them. Finally, we included questions relating to financial performance to help with an industry profile that PricewaterhouseCoopers was interested in developing.

These three objectives were combined into research instruments that were piloted and revised, then tested again in a fax-out version of the survey. The pilots were conducted by student researchers who called companies listed in the database, asked them to participate, and then sent them a survey by fax. The results of these pilot tests were not encouraging. Response rate was very low, which suggests that detailed surveys are difficult to formulate in a way that can ensure broad participation without extensive follow-up. Questions relating to financial performance were especially touchy topics, and many respondents left these questions blank. Questions relating to innovation and knowledge management seemed too detailed for many firms. The small companies often did not have these activities specified or managed separately from the day-to-day management of the organization.

The survey was also piloted in face-to-face interviews with a group of students doing projects on the new media industry. These students used the survey as a framework for in-depth interviews with approximately 30 companies in the new media industry. Once students were able to get an appointment with the appropriate person within a company – an often difficult task – they encountered resistance to answering financial questions and a relative lack of comprehension of the innovation and knowledge management questions.

Based on the feedback from the students doing interviews, and from those who conducted the call-outs to solicit participation in the fax-back version of the survey, we again revised and simplified the survey sheets.

We also did some testing of the online survey forms. Anecdotal evidence from student interviewers suggests that in some parts of the high technology sector, most notably at the software and Internet firms, people are interested in and willing to use an online form. Others (e.g., in manufacturing) are not willing to or interested in doing so.

An abbreviated and simplified survey instrument, as well as a guide for face-to-face and telephone interviews, was finally adopted and used.

Data collection and management system (Objective 2)

Database design. The data for Techmap II is contained in a number of related databases. These databases were constructed using mySQL and housed on a UNIX server at CPROST. Data is accessed either directly through a console (command-line) interface to mySQL, or through MS Access or Filemaker using OBDC. Access to the data is also available via the worldwide web using a web-based server-side front end.

The database contains tombstone information in one table: Company name, address, business type, contact person, etc. A unique identifier tags each record in this table. Three tables – corporate performance, innovation survey results, and company history – contain the basic survey data related by firm identifier. Another table contains the results of the knowledge management section of the survey.

In addition, two linkage tables examine the relationships between the firms. In the first, each record represents collaboration between two firms. The record contains the identifiers for the two firms, and information about the type of relationship: its date, duration, and result. In the second table, each record represents an employee.[3] Each record contains the last five employers, the dates of employment at each, and the educational record of the employee. A third linkage table, genealogies, is planned. This table will track changes in ownership and flows of investment capital.

Overall design, management and growth (Objective 3)

Collecting the initial data. Several assumptions were made in starting the Techmap II database:

[3] Employees are anonymous. No record is kept that in any way identifies individuals.

- A separate genealogy database would be created containing firms no longer in operation. We also assumed that the original Techmap was fairly complete in enumerating these firms.
- We were primarily interested in firms that create or add knowledge to products or services in B.C. Retailers, manufacturers' reps, distributors, etc., would not be included in the database.
- As much as possible, a standard industrial classification scheme would be used. It was decided to use the NAICS (North America Industrial Classification System) scheme, since conversions to NAICS from other schemes are available. (This proved to have been an incorrect assumption at the time: Statistics Canada's literature stated that a concordance was available in June 1998, but it had still not been released by December 1998).
- Firms of fewer than 5 employees would not be included in the database, unless they had been in existence for more than 5 years.

There are several large commercial databases, such as Dominion Directories and Contacts Target Marketing, that contain this type of information. These databases were not used as sources of data for the Techmap II database; rather they were used to confirm and verify data from other sources.
Most of the entries in the database contain at least the following information: name, address, phone, fax, e-mail, industrial classification (SIC80, SIC87, or NAICS), an estimate of the number of employees, and a contact. There are currently a total of 1440 firms in the Techmap II database as of February 2000.

Initial techmap database. 439 firms were included in the first Techmap database. Of these, 318 were imported into the Techmap II database. The 121 firms not imported were no longer in operation.

Classification of firms by industrial activity. Based on the Techmap I data, and on the BC Stats definition of high technology, working definitions of high tech according to the NAICS, SIC80 and SIC87 classification schemes were established. This allowed relatively easy filtering of firms from other databases, assuming they were classified using one of these schemes.

B.C. new media. A contact list of 320 names was received from New MediaBC, an industry association. Fourteen of these firms were in the Techmap I database. The list was screened through web searches against Dominion Directories and resulted in 200 firms from this list being imported into the Techmap II database.

After this step, the Techmap II database contained 518 entries.

CPROST databases. Databases from previous CPROST research were merged with the Techmap II database, yielding 188 firms not previously included. After this step, the Techmap II database contained 706 entries

B.C. manufacturers directory. Using the definition of High Tech established by B.C. Statistics, a list of firms was extracted from the 1997 B.C. Manufacturers Directory, containing the names of about 700 firms. This list was merged with the Techmap II list, adding 379 firms for a total of 1085 entries.

Business in Vancouver high tech directory. The data in the BIV directory is collected through firms asking to be included on the list, and is therefore somewhat suspect. The 1998 list contains 1534 high tech firms, classified using an internal scheme. The first step in importing this data was to build a concordance with the NAICS scheme, allowing firms to be sorted by industrial activity. The BIV list was then purged of retailers and wholesalers, firms under 5 employees, firms in classifications not included in the Techmap definition of high tech, and firms already in the Techmap II database. Of the original 1534 names in this list, 320 were added to the Techmap II database, for a total of 1405 entries.

Other data sources. Some data has been received from other sources, including the Science Council of British Columbia. Data based on firms discovered through media reports, web searches, and personal networking, has also been entered by the researchers.

Summary, Conclusion and Recommendations

The three objectives of the research program (design and revise a research methodology, build and test a data collection system, and build the database of firms) were complete by December 1998. In addition, data from the pilot interviews and fax-back surveys completed in the fall of 1998 has been collected from about 50 companies. This data has not been entered into the new database system but could be moved over.

The methodology has been tested and revised significantly, but there remain concerns that the survey is still onerous to complete. Significant field and follow-up resources will be required if we continue with this approach. Although we had hoped that the data collected by students in their projects would move us much closer to Techmap II, it is clear that a smaller number of trained full-time personnel, and a less intrusive questionnaire, represent a better approach than a large number of part-time researchers. Training should include some time spent accompanying a person familiar with the process of collecting Techmap data.

The online data collection system needs further refinement to complctc thc automation of the firm linkages process and validate all of the data insertion codes. Once completed, links to the ASI and PWC web sites and advertisements about the Techmap should direct people to the Techmap web site and encourage them to complete a survey on their own.

Unfortunately, the genealogical representation of Techmap data can only take us so far. At the suggestion of one of the earlier reviewers of the this paper, the author has begun to investigate the potential contribution that could be made to this work through the application of graph theory (see West 1996) for an introduction to the concepts and uses of graph theory).

Finally, it is important to recognize that such maps are – like most forms of academic research – mere "snapshots" of reality. In the case of the technology industry, this "freeze-frame" aspect is particularly noticeable because of the highly dynamic nature of the sector. Our self-imposed restriction to firms that had been in business more than five years or had more than five employees also had the effect of reducing some of the currency of the map. Nevertheless, "face validity" in the process and the results can be found in the presence of the map in numerous offices and boardrooms around British Columbia, even four years after it was created.

The Future

Online data representation has long been a concern for us. Even before the addition of a much larger number of firms, we had difficulty representing our results on paper. With the second version of the Techmap, we were proposing to collect a much richer set of information about the firms. How do we represent information about 1400 firms on a map that was crowded at 400? In addition, how can we account for the movement of firms in and out of the geographic area covered by the Techmap? Clearly firms don't necessarily cease to exist if they move their offices.

This is a particular concern when accounting for brain drain and brain gain and the movement of head offices. Some form of notation that provides information about where firms come from and go to would be helpful in this regard. This need for alternative views of the map highlights the necessity of obtaining multiple perspectives on the data and matching representational and analytical tools to the questions at hand. While the genealogical portrayal was sufficient for our initial, descriptive, approach, more fine-grained and powerful tools will have to be developed for further studies in more targeted areas.

The obvious solution to the problem of the growing number of firms in the database was a web-based representation, making use of the hypertext and graphic capabilities inherent in the worldwide web. One of the exciting possibilities is contextual zoom, an emerging technology developed by another group at Simon Fraser University. The software, Czweb, has recently been licensed to a software company in Vancouver, and we hope to use it in future versions of the Techmap. Contextual zoom technologies allow the viewer to zoom in on specific items in a large data set without losing the context of their search. Nearby items shrink and move out of the way on the screen but do not disappear entirely, as they do in hypertext links.

The other software technology that we have been exploring is the use of geographical information systems (GIS) to enable more location-specific mapping of technology firms. In a preliminary test using data from the first Techmap, we were able to quickly locate our firms by latitude and longitude using a program that matches postal codes with physical locations. We were able to do this through the assistance of Ken Jones, a colleague who holds the Eaton Chair in Retailing at the Centre for Studies in Retail Trade at Ryerson and the University of Western Ontario.

Some combination of geographical and contextual location will be used in the final form of the next generation of the Techmap. In addition, we will endeavour to link our relational database to the online (web-based) data representation system. In practical terms, this means that the map, with all its boxes and lines, will be redrawn automatically, based on changes in the database. As it turns out, advances in computer-aided design (CAD) have been headed in a similar direction for many years. Although adapting the tools to our needs is no small task, many of the tools we require have already been developed in GIS and CAD software and in the fields of resource management and architectural and engineering drawings.

Finally, the collection and use of social capital information has begun to seriously impact the direction of the Techmap project. We have begun to focus more closely on the use of case studies, necessary to draw out the subtleties of networks, trust relationships, and norms, in the collection of information for this work. The future direction of the Techmap may very well be away from a map and toward an interactive and hyperlinked data representation that more closely captures the dynamic, fluid reality of interpersonal, interfirm, and industry-wide relationships. To move in that direction, we will have to continue our efforts to refine the data collection and data representation tools we have at our disposal. With this and subsequent rounds of Techmap research, we hope to be able to examine in more detail the linkages between firms, both past and present. In this way we hope to be able to understand the complex networks of relationships within the B.C. high tech community. With better understanding comes the possibility of focused and effective policy responses.[4]

Acknowledgments

This paper has benefited greatly from the assistance of Lindsay Hughes, who helped with the management of the Techmap II research project.

References

Acs, Z., de la Mothe, J., and Paquet, G. (1996). Local systems of innovation: In search of an enabling strategy. In *The Implications of Knowledge-based Growth for Micro-economic Policies*, ed. P. Howett, 339-60. Calgary: University of Calgary Press.

Amidon, D. M. (1997). *Innovation Strategy for the Knowledge Economy: The Ken Awakening*. Boston: Butterworth-Heiman.

Bourdieu, P. (1986). The forms of capital. In, *Handbook of theory and research for the sociology of education*, ed. J. G. Richardson, 241-58. New York: Greenwood.

Chiesa, V., Coughlan, P., and Voss, C. A. (1996). Development of a technical innovation audit. *Journal of Product Innovation Management* 13: 105-36.

Fountain, J. E. (1998). Social capital: its relationship to innovation in science and technology. *Science and public policy* 25(2): 103-15.

Granovetter, M. S. (1973). The strength of weak ties. *American Journal of Sociology* 78(6): 1360-80.

[4] Images to complement this chapter are available at the following website: **http://techmap.cprost.sfu.ca**.

Lundvall, B.A., ed. (1992). *National systems of innovation*. London: Pinter.

Nahapiet, J., and Shoshal, S. (1998). Social capital, intellectual capital and the organizational advantage. *Academy of Management Review* 23(2): 242-66.

Smith, R., Takanashi, A., and Ahmed, M. U. (1998). Dancing with giants – collaboration for technological change in the 21st century. *International Journal for Technology Management*, 17 (3/4): 285-92.

Smith, R., M. Cheung, et al. (1997). TechMap. Vancouver: PricewaterhouseCoopers.

West, D. B. (1996). *Introduction to Graph Theory*. Upper Saddle River, New Jersey: Prentice Hall.

Chapter 4

The Alberta Innovation System: The Main Players and Suggested Research Priorities

Peter Josty

Abstract

Over the last few decades, the study of the innovation process has emerged as an accepted field of multidisciplinary research. This paper maps the institutions that make up the Alberta innovation system using a framework developed by the Organization for Economic Cooperation and Development (OECD) that considers seven key dimensions. These dimensions are: (1) makers of general policy, regulations and taxes; (2) innovators; (3) funders of innovation; (4) generators of new knowledge; (5) funders of new knowledge generation; (6) providers of services related to innovation; and (7) providers of education and training. We identified over 100 separate institutions altogether, and this number would have been much larger if we had considered individual banks, universities, etc. The innovation system is examined by posing four basic questions: What results are expected? How are we doing? Where are the big gaps? How can research on innovation make a contribution?

The answer to this last question proposes ten research priorities for studying innovation in Alberta. These are: (1) better understanding of how the various elements of the innovation system communicate and work together; (2) better understanding of how Alberta-based companies adopt business practices that encourage innovation; (3) better understanding of the processes of innovation in commodity industries; (4) better understanding of how public attitudes and perception about science, technology and innovation influence innovation in Alberta; (5) expanding the number of well-documented case studies of innovations carried out in Alberta; (6) better understanding of the dynamics of new firm creation and evolution; (7) better understanding of the dynamics of industry cluster formation and evolution in Alberta; (8) tracking the diffusion of innovations within established industries and communities in Alberta; (9) finding better methodologies to measure innovation; and (10) better understanding of how major research funds create economic value.

Introduction

Over the last few decades, research on the innovation process has emerged as an accepted field of multidisciplinary academic research. Numerous institutions have sprung up around the world to study innovation; among the first of these was the Science Policy Research Unit (SPRU), founded in 1966 at the University of Sussex, in the U.K.

Some of the major characteristics of innovation have become well established and thoroughly documented (Utterback 1996; Drucker 1985) As the volume of research has grown and understanding of the innovation process has increased, a new strand of research has emerged that looks at the innovation process from a holistic, systemic perspective. Thinking about innovation from a systems perspective allows us to see innovations as just one part of a complex adaptive system (Casti, 1994). The benefits of this approach are that it:

- forces a more comprehensive view of the large number of factors involved in innovation;
- focuses attention on the *interaction* between all the elements of the innovation system;
- offers the promise of developing new modeling and simulation tools to model the innovation process;
- provides a framework for policy makers to determine what interventions produce the best policy results, and to balance the legitimate interests of different groups;
- utilizes the existing body of knowledge on the study of systems, such as systems thinking, system archetypes, system dynamic modeling (Senge, 1990).

This paper will review the concept of a national innovation system, describe the main players in the Alberta innovation system, draw some tentative conclusions, and propose some research priorities to deepen understanding of the functioning of the system. We are exploring the possibilities of establishing this kind of information on a web site for ease of access and updating capabilities.

The Concept of a National Innovation System (NIS)

In 1999 the OECD published "*Managing National Innovation Systems,*" which reported on the work of a task force of OECD member countries. In this report, national innovation systems are defined by Metcalf as the

> set of distinct institutions which jointly and individually contribute to the development and diffusion of new technologies and which provide the framework within which governments form and implement policies to influence the innovation process. As such it is a system of interconnected institutions to create, store and transfer knowledge, skills and artefacts which define new technology. (Metcalf 1995, 409-512)

From this perspective, the innovative performance of an economy depends not only on how the individual institutions (e.g., firms, research institutes, universities) perform in isolation, but on how they interact with each other as elements of a collective system of knowledge creation and use, and on their interplay with social institutions (such as values, norms, legal frameworks) (Smith, 1996).

Why Study the Alberta Innovation System?

It is legitimate to ask at the beginning of a study of innovation in Alberta why this is a worthwhile pursuit. We advance three main reasons for believing this study to be worthwhile:

1. Innovation is becoming much more important to the Alberta economy, and the move toward a knowledge-based economy is changing the quality and nature of the activity. A better understanding of how this process works may provide tools and techniques to allow players to innovate more effectively.
2. Innovation is a localized phenomenon, and can only be understood in terms of the specifics of a particular location. Lessons learned elsewhere may be hard or impossible to apply in Alberta (e.g., Saxenian 1996). However, a

contrary view is that innovation is a global phenomenon only. (See comments by Phillips in Chapter 2 of this book).

3. New tools and concepts are becoming available that may allow for new insights and understanding of the innovation process. Viewing innovation from a systems perspective permits the use of a wide variety of tools and concepts such as systems thinking, system dynamic modeling, system archetypes, fractal behaviours, self-organization, self-similarity and emergence.

The Alberta Innovation System

The OECD study on National Innovation Systems (OECD 1999) describes the innovation systems in a number of countries (France, Sweden, Finland, Austria, Spain, Germany, and Belgium) and uses a variety of different formats to portray those innovation systems. Of the various formats, the one used to describe Sweden's innovation appears most suitable for describing the Alberta system, as it is structured according to the functions of the different institutions, and not the institutions themselves, which differ considerably from country to country.

Table 1: The main players in the Alberta Innovation System

Makers of general policy, regulations and taxes Federal Government Provincial Government Municipal Government Other regulatory bodies International bodies **Innovators** Established organizations Individuals Entrepreneurs Start-up companies **Funders of innovation** Established companies Venture capitalists Angel investors Stock market investors Others	**Generators of new knowledge** Companies Federal agencies Provincial agencies Other research centres Universities **Funders of new knowledge generation** Companies Federal Government Provincial Government Private not for profit foundations Other **Providers of services related to innovation** Federal agencies Provincial agencies	Private service companies Professional service providers Providers of Awards and prizes Media Incubators Innovation Centres Universities **Providers of education and training** Universities Private colleges Technical Institutes Public colleges Vocational schools and colleges Apprenticeships

Table 1 illustrates the structure and main players of the Alberta innovation system as it exists in early 2000, using the Swedish format, slightly modified. Each sector of the innovation system is described under two headings:

- ***The Players*** – who they are and what they do.
- ***Discussion***

In some cases an institution appears twice, as it plays two roles. For example, the National Research Council appears as both a generator of new knowledge and a funder of new knowledge.

In preparing a list like this, there is the question of where to draw the boundary. The basic criterion we have used is an estimate of the influence of an organization on innovation in Alberta. We have taken the view that some influential elements of the Alberta innovation system, particularly those connected with the federal government, are located outside Alberta. Many institutions have been omitted from this list, but an attempt has been made to include the most influential ones. In approaching the boundary issue (what to include, and what to leave out), we have taken a broad view of what influences innovation in Alberta, and therefore of what should be considered part of the innovation system. The author would be glad to discuss adding other organizations to this listing.

General Policy, Regulations and Taxes

In the general policy area, the primary actors are the provincial federal, and, to a lesser extent, municipal governments (Tables 2–4). Their most important role related to innovation is to provide protection for intellectual property rights, patents, copyrights, trademarks, and industrial design and trade secrets (Baldwin 1997).

Another important role they play is to create a positive business climate: political stability, reasonable taxes, respect for the legal system and predictable policies.

Table 2: Federal government institutions

Institution	Mission/purpose	Comments
Canadian Radio-telecommunications Commission (CRTC)	An independent public authority in charge of regulating and supervising Canadian broadcasting and telecommunications.	Regulates broadcasting and telecommunications.
Environment Canada	A science-based government department whose business is helping Canadians live and prosper in an environment that is properly protected and conserved.	Environmental regulation
Prime Minister's Advisory Council on Science & Technology	To provide the Prime Minister with expert, non-partisan advice on national science and technology goals and policies and their application to the Canadian economy.	Established 1996. It has focused in two areas: skills and commercializing university research
Transport Canada	To develop and administer policies, regulations and services for the best possible transportation system.	Transportation regulation.

Table 3: Provincial government institutions

Institution	Mission/purpose	Comments
Alberta Science and Research Authority (ASRA)	To stimulate science and research and encourage a strong economy by recommending policies and priorities for science and research that more effectively utilize the government's investment in this area.	Main advisor to the Provincial government on science & technology. A part of the Ministry of Innovation & Science.
Ministry of Innovation & Science	Main tasks: to ensure Alberta's innovative and scientific community stays strong. to make Alberta a magnet for national & international investment	Alberta Innovation and Science consolidates all government technology, research and science activities into one unique department within the Government of Alberta.

Table 4: Municipal government institutions

Institution	Mission/purpose	Comments
City of Calgary	Sets taxes and municipal regulations	Encourages economic development.
City of Edmonton	Sets taxes and municipal regulations	Encourages economic development.

Discussion: General policy, regulations and taxes

Compared to many jurisdictions round the world, Alberta has a positive and supportive climate for innovation. According to a survey carried out by the Heritage Foundation and the Wall Street Journal, Canada is ranked 11th of 161 countries in terms of the freedom of the economy (Globe and Mail 1999).

However, because of our geography, most comparisons are made with the U.S., which is probably the most innovative economy in the world at the moment. In this comparison, Alberta lacks certain key assets: a critical mass of entreprencurial businesses as found, for example, in Silicon Valley; a highly developed venture capital market; and sufficient numbers of experienced entrepreneurs. Taxes are higher in Alberta, and this cannot be overlooked, due to the potential for migration of people and businesses across the border. However, some of the most innovative economies in the world have high taxes, for example, Finland and Norway, so high taxes alone do not seem to be obstacle to innovation. In Canada, the R&D tax credit provides a benefit to companies who perform R&D. However, Alberta does not have a provincial R&D tax credit to combine with the federal R&D tax credit.

Innovators

Innovators are the individuals and small groups who drive the innovation process. Ultimately, all the institutions described in this paper exist to make this group more effective. Innovators provide the energy and drive to take an idea from the stage where it exists in a laboratory or someone's mind and drive it toward commercialization. These individuals are the essential core of the entrepreneurial economy. Innovation may also exist in universities, governments, and other non-business settings. In large established organizations, many barriers exist to innovation, particularly to radical innovations (Freeman and Soete 1997).

Bhidé classifies new business initiatives into four categories (Bhidé 2000):

- *Bootstrap start-ups*. These start-ups represent the vast majority of new business incorporations. Many of these businesses make no pretense at being innovative (e.g., a hairdresser, new retail outlet, a franchise), and are usually funded by the owner, and by family and friends. Generally there is very little planning, and the focus is on opportunistic adaptation with expertise provided by the owner. Most do not grow significantly. Some very successful companies start in this manner (e.g., Microsoft, Hewlett Packard), but their early growth is very slow.

- *Venture capital based start-ups.* Venture capitalists provide capital to an elite group of entrepreneurs who have innovative ideas or technologies and a verifiable track record of business or technical achievement. These start-ups have strong plans, and usually utilize expertise from the venture capital firm. Growth tends to be very strong, as the venture capitalist needs to recover the investment within a three to five year period. These start-ups face extensive scrutiny from the providers of capital. Examples of venture capital based start-ups include Compaq and Lotus.
- *Corporate initiatives.* Corporate initiatives tend to focus on large opportunities that require significant resources and have low uncertainty. Companies have extensive control systems to protect stockholders' interests, and these usually demand that decisions be made in a systematic and thorough manner. Success in a corporate venture is usually determined by a team rather than by a brilliant individual.
- *Revolutionary ventures.* A tiny number of exceptional new business initiatives fall into this category. These are the ventures that bring revolutionary ideas to market (e.g., Federal Express and Motorola's Iridium project). These projects have a systemic aspect: they cannot start small and grow. Their very nature requires them to be big. They are founded by extraordinary individuals.

Discussion: Innovators

Much of the focus of the institutions in the Alberta innovation system is on the small company start-up. There is a good reason for this: large corporations generally have the resources to be self-sufficient in this regard. Paradoxically, much less is known about the process of innovation in start-ups and small companies than in established companies. Most business research and most consulting companies focus on large companies, as the data is much more available and the consulting fees are larger (Bhidé 2000).

There are more than 200,000 incorporated businesses in Alberta, 75% of which employ five people or less, and 95% of which employ fewer than 50 (Frank 2000). Most new jobs are created by small

Table 5: Funders of innovation

Source	Information	Comments
Agrivest Capital Corporation	A venture capital firm based in High River, AB, founded in 1996	Minimum investment $100,000
Alberta Value added Corporation (AVAC)	Capital of $35 million.	A private company with start up funding from Ottawa and Edmonton to support growth of value added agricultural products in Alberta.
Capital markets	$1.1 billion was raised by companies listed on the Alberta Stock Exchange in 1998	Not all of this money was raised for innovation purposes.
Companies retained earnings	Size uncertain.	Includes sweat equity investments by entrepreneurs.
Federal Business Development Bank	To help create and grow Canadian small- and medium-sized businesses through a range of financial and management services	Founded in the 1940's. Mandate revised in 1995. Has a venture capital arm.
Industrial Research Assistance program (IRAP)	For more than 50 years, we've been helping small businesses turn good ideas into commercially viable products and services.	Part of NRC. Founded in 1945. Provides funding support for early stage commercialization.
Launchworks	A Canadian incubator of emerging technology companies founded 1999 in Calgary.	A major US venture capital company – Inter-Ventures- announced the acquisition of a 31% stake in Launchworks in April 2000.
Manvest Inc.	Private Calgary based venture capital company, founded in 1978.	Prefers deals in the $2-5 million range.
Tamarack Group	Venture capital company in Calgary founded 1991.	Prefers deals in the $1-3 million range.
The Alberta Opportunity Company	AOC was established to help Albertans turn their good ideas into thriving businesses - by providing loans when conventional financing is not available.	Established 1972
Technology Partnership Canada	Flagship Canadian Government fund for industrial support.	$300 million a year. Most goes to larger companies, but an increasing amount is spent on small technology start up companies.
Venture Alberta.com	A match making service bringing together investors and entrepreneurs.	Said to have access to $350 million in venture capital.
Venture capitalists	$135 million invested in Alberta in 1999	

Table 6: Venture capital investments in Canada (millions of dollars)

Year	1994	1995	1996	1997	1998	1999
$ invested	$459	$669	$1094	$1822	$1656	$2720

companies (Birch 1979). It is necessary to distinguish between entrepreneurs and innovators. We see innovators as the subset of entrepreneurs who pursue innovative opportunities.

Funders of Innovation

Funders of innovation provide the risk capital to fund innovative projects, which are usually seen as having high business risk (Table 5).

Discussion: Funders of innovation

Venture capital

According to MacDonald (2000), venture capital investments in Alberta in 1999 totalled $135 million, up 45% from 1998. This represents 5% of total Canadian Venture capital investments. The average size of investment was $4.6 million, up from $2.0 million in 1998.

In Canada as a whole, the total pool of venture capital at the end of 1999 was $12.1 billion. This figure has more than doubled from $5 billion in 1994. Venture capital investments in Canada have risen sharply in recent years, as shown in Table 6 above.

It should be noted that venture capital tends to enter at a later phase of the innovation process.

Capital markets

In 1998, companies listed on the Alberta Stock Exchange (ASE) raised $1.1 billion in fresh capital, compared to $1.4 billion in 1997. There were 151 new companies listed on the ASE.

From companies themselves

Unfortunately, statistics on innovation spending (marketing costs, significant organizational changes, improvement in business processes, some capital costs, etc.) are not easy to obtain. R&D spending has been tracked for many years, and statistics are readily available. This situation tends to attract too much attention to the R&D spending statistics, at the expense of the broader picture.

The banks

According to the Canadian Bankers Association, banks are the largest source of financing for small businesses, but they are not the only source. Banks provide just over 50% of debt financing to small and medium-sized enterprises in Canada. The other 50% comes from other providers such as credit unions, caisses populaires, trust companies, specialized finance companies, and government lending agencies.

Government

The Technology Partnership Canada fund has been criticized for its investment in small companies, a field felt better left to venture capitalists (Reguly 2000).

Generators of New Knowledge

New knowledge generation is one of the most important parts of the innovation system and is changing significantly. In the past, new knowledge was generated slowly by a process of trial and error based on experience (Landes 1998). Today, new knowledge is generated and communicated through very systematic processes in R&D laboratories and other places and, as a result, is produced much more quickly that it was in the past.

New knowledge can be exploited by innovators to create new commercial products and processes to drive economic growth. The relationship between knowledge production and wealth creation should not be seen as a linear process, but as a process involving complex feedback loops, time delays and many other factors.

Discussion: Generators of new knowledge

There is a rich diversity of institutions in Alberta producing new knowledge (Table 7). There appears to be a good balance between those that focus on the new economy areas, such as biotechnology and electronics, and on the traditional economy areas, such as agriculture, petroleum and forestry.

Funders of New Knowledge Generation

The funders and the generators of new knowledge are frequently different organizations (Table 8). Close to $1 billion is spent in Alberta each year on generating new knowledge. Roughly half is spent by the public sector, and half by the private sector. The overall level of spending in Alberta is less than the Canadian average and much less than in some other countries.

Table 7: Generators of new knowledge in Alberta

Institution	Mission/purpose	Comments
Alberta Agricultural Research Institute (AARI)	"AARI coordinates, supports and promotes research and technology in the agriculture and food industry." AARI	1998/99 budget of $9.1 million. Founded 1987. 13 administrative personnel.
Alberta Microelectronics Corporation (AMC)	"The AMC is a world leader in the design and fabrication of micro-machined glass, silicon and thin film components as well as microelectronic instrumentation." AMC	Located at U of A campus. Founded 1988. Provides service to companies in AB and round the world.
Alberta Oil Sands Technology and Research Authority (AOSTRA)	"To facilitate research for the economic and sustainable development of oil sands, heavy oil, and enhanced oil recovery in crude oil." AOSTRA	Established 1974. Currently runs COURSE, a new (1999) initiative to increase and focus the amount of fundamental research being conducted at universities in Alberta related to the energy resources industry.
Alberta Research Council (ARC)	"ARC works with customers and partners to bring technology developments into commercial use and provides research, development, and advisory services that effectively bridge the gap between research and market development." ARC	Approximately 600 employees. Founded 1921. Focus in 7 areas: agriculture, biotechnology, energy, environmental, forestry, information technology, and manufacturing. Now includes PRI (Petroleum Research Institute) and C-FER (Centre for Engineering Research).
Canadian Network for the Advancement of Research, Industry & Education (CANARIE)	"Canada's advanced Internet development organization. It works with government, industry, and the research and educational communities to enhance Canada's advanced Internet infrastructure, applications development and use." CANARIE	120 members, and 500 partners. Established 1993. National sponsor for the Netera Alliance.

Institution	Mission/purpose	Comments
Food Processing Development Centre	Conducts research for Alberta's food and beverage industry.	Founded 1984. Funded by AHFMR.
Informatics Circle of Research Excellence (iCore)	To foster university-based long-term fundamental research that supports the information and communications technology (ICT) sector.	Funded for an initial 3-year period by the Alberta Government, $30 million. Founded October 1999
Netera Alliance Inc.	Netera Alliance is a not-for-profit corporation of universities, research institutions, government and small and large private-sector companies facilitating the advanced networking and computing infrastructure for Alberta.	Formerly Wurcnet. Originally formed 1993.
Private companies		Spent $476 on R&D in 1997 in Alberta.
Telecommunication Research Labs (TRLabs)	"TRLabs is Canada's leading research consortium in information and communications technology." Published mission statement.	Funded by governments of Canada, Alberta, Saskatchewan, Manitoba, private funders, and by universities. Annual budget $10 million. Founded 1986.
The Canada Centre for Mineral and Energy Technology (CANMET) Western Research centre	Through partnerships and scientific research, it is the mission of the CANMET Western Research Centre (CWRC) to deliver innovative technology and solutions to energy-related industries for sustainable development.	Focus on: 1. Advanced separation technology. 2. National Centre for upgrading technology. About 130 employees. Located at Devon, AB. Originally founded 1955.
University of Alberta (U of A)		Founded 1908, the U of A is the largest research institute in the province, with 30,000 students & 3800 academic staff, and 3300 support staff.
University of Calgary (U of C)		Full autonomy 1966. 22,000 students, 1,700 academic staff, 2000 support staff.

Table 8: Funders of new knowledge generation in Alberta.

Institution	Mission/purpose	Comments
Alberta Heritage Foundation for Medical Research (AHFMR)	AHFMR supports medical research that improves the health and quality of life of Albertans.	2000 spending likely to exceed $40 million Founded 1980. Original endowment $300 million.
Alberta Heritage Foundation for Science and Engineering Research (AHFSER).	Likely to be modelled on AHFMR.	Announced in February 2000. Planned initial endowment of $500 million with promise of more later.
Alberta Science and Research Authority. (ASRA)	To stimulate science and research and encourage a strong economy by recommending policies and priorities for science and research that more effectively utilize the government's investment in this area.	Founded 1994
ASRA Intellectual Infrastructure Partnership Program (IIPP)	To increase research excellence and competitiveness of Alberta universities and research hospitals through investing strategically in their research infrastructure.	Matches CFI funding.
ASRA Science & Research Fund	To provide financial support for strategic science and research initiatives that will produce human, social, environmental or economic benefits for Albertans.	Focus on infrastructure.
Canada Foundation for Innovation (CFI)	To increase the capability of Canadian universities, colleges, hospitals, and other not-for-profit institutions to carry out important world-class scientific research and technology development. Established 1997.	CFI funds have been increased from an initial $800 million. Spending will be about $180 million per year.

Institution	Mission/purpose	Comments
Canadian Institute of Health Research (CIHR)	Replacement for the Medical Research Council, founded 2000. Broader mandate than MRC, for biomedical research, health services, applied clinical research, and social, economic and population health.	First year budget is $500 million, and by 2004 it could be $1 billion.
Medical Research Council of Canada (MRC)	The Medical Research Council of Canada (MRC) is the major federal agency responsible for funding biomedical health research in Canada. Its role is to promote, assist and undertake basic biomedical research, applied and clinical research in Canada in the health sciences. In 1998/99 about 12% of the total operating grants were spent in Alberta.	Founded 1969. Funds research 1998/99 budget $267 million. MRC has been replaced as of April 1 2000 by the Canadian Institute of Health Research.
National Science & Engineering Research Council) (NSERC)	NSERC's role is to make investments in people, discovery and innovation for the benefit of all Canadians. Invests in people by supporting more than 9,000 students in their advanced studies. Promotes discovery by funding more than 8,700 researchers every year. Helps make innovation happen by encouraging more than 1,000 Canadian companies to invest in university research.	Spent $42.5 million in Alberta in 1998/99. Founded 1978.
Private companies		Spent $476 million on R&D in 1997 in Alberta.
Research- based pharmaceutical companies	In November 1999, pharmaceutical companies announced spending targets of $150 million in Alberta over the next 3 years on R&D.	$33.2 million invested in Alberta in 1998 in R&D.

Institution	Mission/purpose	Comments
Social Sciences and Humanities Research Council (SSHRC)	SSHRC is Canada's federal funding agency for university-based research and graduate training in the social sciences and humanities.	Budget in 98/99 was $104 million (Canada wide). Established 1977.
Western Economic Diversification (WED)	To promote the development and diversification of the economy of Western Canada and to advance the interests of the West in national economic policy.	Established 1987. 10 staff in Calgary, 85 in Edmonton.

Table 9: Alberta R&D spending 1997

Funder	**$ spent in 1997, in millions**
Federal Government	$196
Provincial Government	$103
Business Enterprises	$476
Higher Education	$126
Foreign	$83
TOTAL	$984

Source: Statistics Canada (1999)

Discussion: Funders of New Knowledge Generation

New knowledge generation tends to be synonymous with research and development, although R&D is not necessarily associated with innovation or entrepreneurship. R&D spending in Alberta in total (GERD) is about 1% of GDP, compared with the Canadian average of 1.6 % (Tables 9 ans 10). This is lower than the levels of R&D spending of our major competitors, such as the U.S. (2.4%) and Japan (2.9%).

Over the last 10 years, R&D spending in the provinces of Alberta has fluctuated around a level of 1% of PGDP (provincial gross domestic product). This contrasts with the steady rise in R&D spending in Canada's two largest provinces — Ontario and Quebec — and in Canada as a whole.

In 1998/99 the Alberta government budgeted $150 million for spending on R&D (Table 11). This figure includes $33 million from the Alberta Heritage Foundation for Medical Research.

Table 10: Trends in R&D spending in Canada 1988-1997.
Dollars in millions.

Year	$ spent in Alberta	Alberta R&D spending as % PGDP	Ontario R&D spending as % of PGDP	Quebec R&D spending as % PGDP	Total Canada R&D spending as % of GDP
1988	$623	1.00	1.52	1.42	1.40
1989	$654	1.00	1.47	1.49	1.39
1990	$763	1.07	1.56	1.61	1.47
1991	$764	1.07	1.62	1.74	1.52
1992	$761	1.02	1.65	1.88	1.56
1993	$819	1.02	1.75	1.93	1.60
1994	$914	1.08	1.85	1.97	1.66
1995	$920	1.05	1.80	2.02	1.64
1996	$919	0.98	1.76	2.05	1.60
1997	$984	0.97	1.80	2.10	1.62

Table 11: Provincial government R&D spending, as reported by ASRA for 1998-99

Department	Total government spending, $millions
Advanced Education & Career Development	18.5
Agriculture, Food & Rural Development	16.37
Alberta Agricultural Research Institute (AARI)	9.10
Economic Development	3.08
Energy	20.30
Environmental protection	2.25
Health	5.75
Alberta Heritage Fund for Medical Research	33.00
Justice	0.32
Alberta Research Council	33.44
(less Department Contracts to ARC)	(8.37)
Alberta Science & Research Authority (ASRA)	16.5
Transportation and Utilities	0.45
TOTAL	**150.69**

Source: ASRA

The combined (federal and provincial) government share of total R&D spending – 31.7% in 1997 — is almost identical to the percentage in the United States (31.6%) and in line with that in other OECD countries (Baldwin 1997).

To some degree, the low R&D spending in Alberta should come as no surprise, given the large commodity companies who dominate the economy today. From an institutional point of view, there are clearly numerous sources of funding for new knowledge generation, so this aspect of the innovation system would appear to be quite healthy.

Providers of Services Related to Innovation

Services related to innovation cover a multitude of areas such as patent protection, training, providing of facilities such as research parks and incubators, networking organizations, tax, legal and financial advice, etc. (Table 12).

Table 12: Providers of services related to innovation in Alberta.

Institution	Mission/purpose	comments
Agriculture & Food Council	To lead and guide Alberta's agriculture and food industry in realizing the vision of Creating Tomorrow.	To increase Alberta's share of world food, agriculture and agribusiness markets. Formed 1993.
Alberta Biotech Association	To provide leadership toward the creation of a strong, viable biotechnology industry.	50 companies active in Alberta. Founded 1998.
Alberta New Media	Represents the growing number of companies and individuals working in emerging technologies within the information, communications and entertainment industries.	Areas such as internet web development, digital graphics and animation, advanced audio, interactive CD-Roms, electronic commerce, electronic gaming. and other dynamic, evolving technologies. Founded 1998.
Alberta Research Council (ARC)	Works with customers and partners to bring technology developments into commercial use and provides research, development, and advisory services that effectively bridge the	

Institution	Mission/purpose	comments
	gap between research and market development.	
BIOTECanada	To foster an environment that responds to the needs of the biotechnology community, both nationally and internationally.	BIOTECanada represents the vast majority of the Canadian biotechnology community, including industry, professional and industrial associations, universities, research centres, and regional groups. Founded 1998. 107 members
Business Development Bank of Canada	To help create and grow Canadian small- and medium-sized businesses through a range of financial and management services.	Founded in the 1940s. Mandate revised in 1995.
Calgary Council of Advanced Technology (CCAT)	Provides a forum to enhance technology awareness, business development and networking for the advanced technology community.	Formed in 1983. About 120 members currently.
Calgary Innovation Centre (CIC)	Assists innovators, businesses and entrepreneurs in fostering and developing their operations.	The CIC acts as a "catalyst" to foster and create new companies with a focus in the life sciences. Part of CRDA.
Calgary Research and Development Authority. (CRDA)	To diversify Calgary's economy by attracting, growing and developing research and knowledge-intensive organizations which produce science and technology intensive products and services, and/or conduct research and development.	Partnership of the City of Calgary, the Calgary Chamber of Commerce, and the University of Calgary. Established 1981.
Calgary Technology Centre	Incubator, part of the CRDA.	Currently 60,000 sq. feet, being expanded by 54,000 sq. feet.

Institution	Mission/purpose	comments
Canadian Advanced Technology Alliance (CATA)	To stimulate "Global Business Growth" through the forces of Canadian innovation and strategic partnership.	1000 members. Formed 1977.
Canadian Information Processing Society	To provide a quality environment where local Calgarians can gather to learn, share and develop resources within the information processing community.	Founded 1958. More than 6,000 IT professional members.
Canadian Intellectual Property Office (CIPO)	• Fostering the use of intellectual property systems and the exploitation of intellectual property information; • Encouraging invention, innovation and creativity in Canada; • Administering the intellectual property systems in Canada • Promoting Canada's international intellectual property interests	Patents, trademarks, copyrights, industrial designs, and integrated circuit topographies.
Canadian Petroleum Institute	Creates value for international clients and the Canadian petroleum industry by delivering Canadian experience and expertise through management and technology training.	Originally founded 1982. An independent, not-for-profit organization.
Canadian Technology Network (CTN)	To provide integrated, accessible pathways to information and services relevant to small and medium-sized enterprises using technology.	Run by NRC. CTN has about 60 members in Alberta and the NWT Links federal and provincial government labs and agencies, universities, community colleges, industry associations, technology centres and economic development agencies. Founded 1995.
Conference Board of Canada	Helps members anticipate and respond to the increasingly changing global economy.	Major private sector player related to innovation. Founded 1954. Currently employs about 200 people.

Institution	Mission/purpose	comments
Economic Development Edmonton (EDE)	Created in 1993 to facilitate and promote a strong, diversified economy and tourism industry in the Edmonton Alberta Capital Region.	
Edmonton Council of Advanced Technology (ECAT)	A forum for discussion for the high tech community.	
Edmonton Innovation Centre	Similar role to the Calgary Innovation Centre.	
Edmonton Research park	320 acres, 35 companies, 1200 scientists.	Established 1980.
Electronics Industry Association of Alberta (EIAA)	Aims to be the single voice for the growing electronics industry in Alberta, business and the public.	All companies, from the largest to the smallest, benefit from joining together for stronger representation of the industry to government.
Industrial Research Assistance Programme (IRAP)	For more than 50 years, IRAP has helped small and medium-sized Canadian enterprises that create and adopt innovative technologies that yield new products, create high quality jobs, and make industry more competitive.	Part of NRC.
Industry Canada (IC)	To foster a growing competitive, knowledge-based Canadian economy.	Current priorities are: innovation, connectedness, marketplace, investment and trade. Runs Strategis web site.
Industry Liaison Office, Univ. of Alberta (ILO)	Provides services in technology transfer and industry liaison activities to aid the University of Alberta in the generation and transfer of knowledge to benefit Canadian society.	Originated in 1985.

Institution	Mission/purpose	comments
Innovation Management Association of Canada (IMAC)	To be the premier Canadian association for leaders in the technology innovation process.	IMAC is an association of research, technology management, and innovation leaders in Canada. About 200 members. Founded 1999.
Licensing Executives Society (LES)	The Licensing Executives Society (U.S.A. & Canada), Inc. is a professional society comprised of over 4,700 members engaged in the transfer, use, development, manufacture and marketing of intellectual Property.	
Management consultants	Provision of management consulting services.	Over 350 companies listed in Calgary alone.
National Research Council (NRC)	To support national science and engineering activities, perform and stimulate investment in research and development, and develop vital expertise and knowledge.	Runs the IRAP ($102 million) and Canadian Technology network programs. 3,000 employees.
Petroleum Technology Alliance of Canada (PTAC)	Provides a structure that brings stakeholders together to identify industry problems and to define research and technology development projects for investigating solutions to these problems.	126 members. Founded 1996.
Promoting Calgary, Inc.	Established in order to coordinate Calgary's economic, business and tourist development initiatives.	Established March 1999
Statistics Canada	Statistics Canada is the country's national statistical agency.	Role goes back to the 17th century.
The TBL Napkin	To provide Founders, CEOs, and Investors of Internet and software technology companies based in Alberta a much needed forum to network, discuss ideas, build	About 70 members TBL are the initials of Tim Berners-Lee, inventor of the World Wide Web,

Institution	Mission/purpose	comments
	business relationships, and even demo your next killer app.	
University Research Park	Incubator.	125 acres in Calgary. Managed by CRDA.
University Technologies Inc. (UTI)	To the pursuit of excellence in technology transfer, creating business opportunities from scientific innovations.	Launched 1989. Serves Univ. of Calgary, Univ. of Lethbridge, CRHA and corporate clients.
Venture Development Group, U of C	Since 1984, our mission has been to promote economic development through entrepreneurial education and research, both nationally and internationally.	The University of Calgary's Venture Development Group is one of the ten best entrepreneurship education programs in North America.
Western Economic Diversification (WED)	To promote the development and diversification of the economy of western Canada and to advance the interests of the West in national economic policy.	Main focus is small and medium-sized companies.

Discussion: Providers of services related to innovation

There are many institutions in Alberta who provide services related to innovation. Many of these target small and medium-sized businesses. There has been a good deal of merging and reconfiguration of these groups in recent years.

Providers of Education and Training

According to the Alberta Ministry of Advanced Education and Career Development, in 1997/98 there were 123,154 individuals enrolled in post-secondary education in Alberta. In addition, there were 27,179 individuals enrolled in apprenticeship programs, and 57,098 enrolled in adult development programs. These were enrolled in the following institutions (Table 13).

Table 13: Providers of Education and Training

Type of institution	Number of institutions in Alberta	Number of full time equivalent students
Universities	5	64,553
Public Colleges	11	29,148
Technical Institutes	2	17,470
Private Colleges	4	2,398
Vocational Colleges	4	9,585
Apprenticeship registrations		27,179
Adult Development programs		57,098

In addition, there are 138 private vocational colleges. On top of this list of formal educational institutions, many companies carry out in-house training programs.

Comparing Alberta with Canada, in 1993 the adult participation rate in all types of credit and non-credit learning activities was 40.1% for Alberta, and 34.8% for Canada.

Discussion: Providers of education and training

Alberta has a solid program of educational offerings from a wide variety of institutions. There are two unique aspects of education in Alberta:

- The emphasis on distance education, in which Alberta is considered a North American leader.
- The entrepreneurial training from the New Venture Group at the University of Calgary, which is one of only a handful of such courses in North America.

According to a survey by the Calgary Research and Development Authority, there were roughly 86,000 high tech jobs in Alberta in 1998, and an estimated 100,000 by the end of 1999. Of these jobs, 54% are in Calgary, 33% in Edmonton, and 13% elsewhere in the province.

A study prepared by Ernst and Young and issued in March 2000 found that more than half of the respondents from high tech firms said it was difficult or very difficult to find adequately trained and experienced R&D and marketing staff.

Overall Trends

Looking at all the elements of the Alberta innovation system, it is possible to identify some trends that appear to be established and will continue into the future.

- **Makers of general policy, regulations and taxes.** The trend toward continued increase in support for innovation-related activities appears likely to persist into the future. Recent examples include actual and planned reduction in taxes, at both federal and provincial levels, and simplification of the R&D tax credit.
- **Innovators.** We anticipate continued growth in employment in the high tech sector, and continued growth in the number of high tech firms.
- **Funding of innovation.** The recent increase in venture capital spending in Alberta appears set to continue.
- **Generators of new knowledge.** The announcement of 2000 Canada Research Chairs across Canada will increase research at universities. Alberta's share of these will be approximately 150-160.
- **Funders of new knowledge generation.** The new Alberta Heritage Foundation for Science and Engineering Research will generate a modest increase in spending for new knowledge generation in Alberta. Apart from this, the trend appears flat.
- **Providers of services related to innovation.** This sector is experiencing considerable flux, with considerable merger and new entrant activity. The trend is likely to be toward an even more comprehensive and complex services sector.
- **Education and training.** There appears to be a trend toward ever higher enrolment in most categories of higher education institution, together with continued shortage of experienced R&D and marketing staff.

Overall Conclusions

When examining a major innovation system such as Alberta's, with hundreds of institutions, it is perhaps appropriate to ask some basic questions at the outset, and to structure a conclusion as responses to these questions. We pose the following basic questions:

- What results are expected?
- How are we doing?
- Where are the big gaps?
- How can research on innovation make a contribution?

What results are expected?

The most recent formulation of expectations comes from the Alberta government's February 8, 2000, News Release (Alberta Government 2000). Specifically, the goal is to ensure that "Alberta is a world leader in innovation, research, development and commercialisation of new ideas."

Before discussing this goal, we should first say that having a provincial government prepared to establish this sort of public policy goal is a huge advantage for those connected with innovation. It is a well known best practice in strategic planning to set "stretch goals," because it is obvious that stretch goals cannot be met by incremental improvements, but force more fundamental, radical approaches.

It is hard not to wholeheartedly agree with this goal. Any criticism would concern the adequacy of the plan, and the social implications of achieving this goal. This issue is discussed in question 4 below.

How are we doing?

There has been substantial progress in Alberta over the last 10-15 years:

- The role of government in this area is now clearly defined in a constructive way (i.e., no longer trying to pick winners, balanced budget, low taxes, public support to major R&D Funds; AHFMR and AHFSER; publicly stated goals on innovation).
- A small high tech community has become established in Alberta, and now amounts to about 7% of provincial GDP.
- The FTA and NAFTA have created access to much larger markets.
- The institutional profile has become much richer as

Table 14: Growth of the Institutions in the Alberta Innovation System

Decade	Number of Institutions founded	Cumulative number of institutions
1950's & earlier	23	23
1960's	4	27
1970's	7	34
1980's	14	48
1990's & 2000	21	69

numerous institutional elements of the innovation system have been established, including Western Economic Diversification Canada (1987), Alberta Microelectronics Corporation (1988), UTI (1989), CANARIE (1993), the Canada Foundation for Innovation (1997), just to name a few.

- The economy as a whole has become much more diversified.
- Much more venture capital has become available.
- The two large universities are ranked first and second in Canada for licensing revenue. Their $US5.1 million of licensing revenue represented 40% of the Canadian total in 1998.

The Alberta innovation system has become much richer and more complex in recent years, as shown in Table 14.

However much remains to be done. This is discussed in the next question.

Where are the Big Gaps?

In thinking about moving forward to achieve the expected results, it is helpful to try to identify the biggest gaps. This is necessarily a somewhat subjective exercise. I will list here the anecdotal views that knowledgeable individuals have expressed over the last few months, some privately, some publicly. They are not listed in priority.

- The Canadian E-business Opportunities Roundtable, an industry lobby group, identified the following gaps in Canada as a whole:
    - Changing income tax provisions on the sale of stock options, to encourage high tech start-ups.

    - Increasing the pool of venture capital through a lower capital gains tax, eliminating cross-border share-for-share mergers, removing impediments to the flow of foreign investment into Canadian venture capital funds, and making Canada more attractive for initial public offerings.

  - The complacent attitude of existing businesses towards e-business
  - The scarcity of e-business clusters in the country.
  - Barriers – particularly a scarcity of capital and higher taxes – that prevent Canadian technology companies from going public.

- Roger Martin, Dean of the Rotman School of Management at the University of Toronto, identified the timidity of the strategy followed by Canadian companies as a prime cause of low levels of innovation in Canada. (Martin 2000). This point was also identified by Finance Minister Paul Martin who spoke about the need for Canadians to embrace a "culture of risk" (Reguly 2000).
- Reports from ASRA have pointed to the low R&D spending in Alberta as a major gap.
- There is a widespread view that the innovation system in Alberta is fragmented and could be more effective. While there are many strong elements in the system, the lack of communication reduces its effectiveness. It has little integrated consciousness as a system.
- Some believe there is a shortage of venture capital in Alberta.
- Others believe there is a severe shortage of entrepreneurs in Alberta.
- One cultural factor mentioned by Roger Martin is the National Policy pursued by successive Canadian governments for over 100 years (from 1879) that sought to make Canada self-sufficient in manufacturing, and to protect manufactures with high tariff barriers. That may have tended to make companies complacent and not aggressive (Martin 2000).

- Alberta lacks the critical mass of knowledge and expertise found in many U.S. and European centres.
- Sherry Cooper, an economist with Nesbitt Burns, believes regulatory barriers should be lowered to enhance innovation (Cooper 2000).

How Can Research on Innovation Contribute?

Given the range of views expressed, there certainly appears to be a need for research to establish some facts in this area, and to help to generate a consensus. The following list represents 10 top research priorities in studying innovation in Alberta. They are not listed in any particular order.

1. **Better understanding of how the various elements of the innovation system communicate and work together.** As noted above, the Alberta innovation system is regarded as very fragmented, and a better understanding of how it functions would allow identification of any gaps or overlaps in the system. In particular, this should include the role of the tax regime, availability of capital, development of entrepreneurs, and the role of regulatory barriers to innovation.
2. **Better understanding of how Alberta-based companies adopt business practices that encourage innovation.** This might include identifying how many companies adopt formal R&D programs, and gaining understanding of the reasons for these decisions. It might also include study of the "innovate in a small start-up companies and then be acquired by established companies" model that is becoming an important mechanism for innovation in software and pharmaceuticals. To what extent does this (or could this?) occur in the oil and gas industry, in the mining industry, etc.?
3. **Better understanding of the processes of innovation in commodity industries**, and the relationship between business strategy, innovation, R&D spending and shareholder value creation. Commodity-producing industries make up by far the largest sector of the Alberta economy today. These industries also innovate, largely but

not exclusively by process innovations. A better understanding of how innovation occurs in these industries would permit better identification and communication of best practices.

4. **Better understanding how public attitudes and perception about science, technology & innovation influence innovation in Alberta.** Many recent pronouncements from the federal and provincial governments have used terminology such as "culture of risk" and "culture of innovation." Public attitudes can strongly influence the propensity to innovate. A recent well-known example is in relation to genetically modified food. Better understanding of the public's attitudes toward various kinds of innovation would help to guide innovation actions by practitioners and governments.

5. **Expanding the number of well-documented case studies of innovations carried out in Alberta.** There have been many successful innovations in Alberta, but this fact does not seem to be widely known or appreciated. Additional case studies and descriptions can provide raw material to inform and influence key audiences and provide further educational material.

6. **Better understanding of the dynamics of new firm creation and evolution,** and the role that technology and the innovation system play. Bhidé (2000) has provided a comprehensive description of this process, based mainly on U.S. examples. It will be worthwhile to validate his description for Alberta.

7. **Better understanding of the dynamics of industry cluster formation and evolution in Alberta,** and the role of the innovation system. Alberta currently has a number of high tech industry clusters, most notably in biotechnology and communications technology. A better understanding of how these clusters have formed and evolved would provide insight into how the present clusters might evolve, and how new ones might become established.

8. **Tracking the diffusion of innovations within established industries and communities in Alberta.** Innovations are constantly being adopted and exploited by the established industries in Alberta. A better knowledge of how this happens might allow identification of best practices and facilitate faster adoption of new technologies, which might enhance productivity growth. Diffusion of innovations is well suited to modeling and simulation, which might allow for further insights into the diffusion process.

9. **Finding better methodologies to measure innovation**. There is an old management adage that "if you can't measure it, you can't manage it." Direct measurement innovation in terms of new products and services is extremely difficult. As a result, most quantification of innovation uses indirect measures such as R&D spending, patents produced, etc. One possible approach is to examine the manufacturing cost in commodity industries as a direct measure of process innovation in those industries.

10. **Better understanding of how major research funds create economic value.** An increasingly popular means of encouraging innovation for governments is to create very large endowments (e.g., the AHFMR, the CFI and the new AHFSER) managed by independent boards. Examining how these funds stimulate new ideas and turn them into innovations may help to create policies that will increase the effectiveness of these funds.

Acknowledgments

I would like to thank Jim Chrisman, Cooper Langford, and Murray Wolfe for helpful comments on earlier drafts of this paper. Any errors are the sole responsibility of the author.

References

Alberta government (2000, February 8). *Get Ready Alberta: Strengthening the Alberta Advantage*. News Release. Edmonton.

Baldwin, J. (1997). *Innovation and Intellectual Property*. Ottawa: Statistics Canada.

Bhidé, A.V. (2000). *The Origins and Evolution of New Business*. New York: Oxford University Press.

Birch, D.L. (1979). *The Job Creation Process*. Cambridge: MIT Program on Neighbourhood and Regional Change.

Canada ranked number eleven (1999, December 1). *Globe and Mail*: B12.

Casti, J.L. (1994). *Complexification*. New York: HarperCollins.

Cooper, S. (2000, February 22). Canada needs a dot-com future: economist. *Globe and Mail*: B1.

Drucker, P.F. (1985). *Innovation and Entrepreneurship. Practice and Principles*. New York: Harper and Row.

Freeman, C. and Soete, L. (1997). *The Economics of Industrial Innovation*. Cambridge: MIT Press.

Frank, C. (2000, February 24). CFIB leader joins small business ranks. *Calgary Herald*: E1.

Landes, D.S. (1998). *The Wealth & Poverty of Nations*. New York: Norton.

MacDonald, M. (2000). 1999 Venture Capital Industry Review and Statistical Tables. (Web page) http://www.canadavc.com/cfm/index.cfm

Martin, R. (2000, January 31). A great crew, but the ship's sinking. *Globe and Mail*: A13.

Metcalf, S. (1995). The Economic Foundations of Technology Policy: Equilibrium and Evolutionary Perspectives. In *Handbook of the Economics of Innovation and Technical Change*, ed. P. Stoneman, 409-512. London: Blackwell.

Organization for Economic Cooperation and Development (OCED) (1999). *Managing National Innovation Systems*. Paris.

Phillips, P.W.B. (2002). Regional systems of innovation as modern entrepots: The case of the Saskatoon biotechnology cluster. In *Innovation and Entrepreneurship in Western Canada: From Family Businesses to Multinationals*, ed. J.J. Chrisman, J.A.D. Holbrook, and J.H. Chua . Calgary: University of Calgary Press.

Reguly, E. (2000, March 4). Martin isn't doing enough to make Canadians risk takers. *Globe and Mail*.

Reguly, E. (2000, March 25). Leave the funding of tech firms to venture capitalists. *Globe and Mail*: B10.

Saxenian, A. (1996). *Regional Advantage: Culture and competition in Silicon Valley and Route 128*. Cambridge: Harvard University Press.

Senge, P.M. (1990). *The Fifth Discipline*. New York: Doubleday.

Smith, K (1996). The Systems Challenge to Innovation Policy. In *Industrie und Glueck. Paradigmenweschel in der Industrie und Technologiepolitik*, ed. W. Polt and B. Weber.Vienna: unknown.

Statistics Canada. (1999). *Science Statistics*, Catalogue 88-001-XIB/ ISSN 1209-1278 November, Vol. 23 No. 6. Ottawa: Statistics Canada.

Utterback, J.M. (1996). *Mastering the Dynamics of Innovation*. Cambridge: Harvard Business School Press.

Chapter 5

Cooperative Mercantilism: Public Policy Dilemmas in Managing the Generation and Application of Ideas Toward Economic Development

Mike Chiasson

Abstract

Ideas have become the essence of innovation, and of business and economic activity. In response to this rise in prominence, increasing concern and confusion have arisen about how to deal with the value, power, and intangibility of ideas in regional economic development. Public policies have largely moved in two directions: (1) neo-mercantilism, which focuses on how to protect the use of ideas, and (2) laissez-faire, which focuses on allowing a barrier-free flow of ideas. Both approaches incorporate serious shortcomings with respect to regional economic development. This paper argues for a cooperative mercantilist approach that encourages widespread exchange of ideas in a way that also permits the development of regional advantages. Case studies and discussions are used to highlight the nature of this approach.

Introduction

In the late 19th century, the economic wisdom of the day conceived of wealth solely in terms of land, labor and capital. Ideas emanating from science, technology and production processes were considered valuable, but difficult to control and largely unacknowledged, and were therefore freely allowed to cross between individuals, organizations and borders (Romer, Basu, Selowsky, and Srinivasan 1992).

Today, ideas have become the essence of economic activity. In response to this rise in their prominence, there has been increasing concern and confusion about how to deal with the value, power and intangibility of ideas in the context of regional economic development (Muniagurria 1995; Pinch and Henry 1999). The plethora of terms that refer to different characteristics of ideas, such as science, technology, innovation, patents, and copyrights, to name only a few, adds to the confusion. Much recent effort has been expended in trying to directly manage the use and development of ideas for economic development. The responses to the need to managing ideas for regional economic advantage vary widely. In particular, public policy has generally focused on one of two diametrically opposed directions: (1) extreme protectionism of ideas to hoard and protect the ideas for regional economic benefit, and (2) a laissez-faire approach, as if intervention is either completely impossible (pragmatic) or completely undesirable (normative) because of the intangible quality of ideas.

The first approach, what I call the neo-mercantilism and protectionism approach, focuses on measures to protect the knowledge and use of ideas that contribute to value-added products and services, at the expense of other regions (Phillips, Chapter 2 of this book; Reich 1992). Those with this protected know-how will be able to produce value-added products to feed the core region's value-added productive capabilities. Patents, copyrights, trademarks, and trade secrets are used as instruments to protect and objectify ideas so that they may be sold in a market. The neo-mercantilist perspective is based on particular assumptions about the relationship of ideas and economic activity: (1) that ideas are commodities that can be developed mostly in isolation from other ideas, and (2) that they can be employed exclusively in the production of local goods and services for export to other regions. As will be discussed in this section, the interconnective value of ideas, the ability of various groups to use a single idea at the same time with little or no added cost, and the discrepancies between idea-producing and idea-using strategies pose some fatal challenges to the protectionist approach (Pasco 1998). This governmental approach is still largely present, but disguised behind the language of free trade (Hagelstam 1991).

The second approach, what I term the laissez-faire strategy, argues against any public policy that restricts the natural flow of ideas among regions and groups. The assumptions underpinning this approach are that ideas cannot be protected and hoarded (pragmatic) and that their value resides solely in their connections with other ideas. Although it addresses some of the shortcomings of the neo-mercantilistic approach, the laissez-faire approach raises other problems. First, ideas do have value when transferred and used elsewhere – and many ideas are developed, at considerable expense, in both the private and public sphere. Second, the laissez-faire approach fails to capture the public policy interventions that have managed to provide important social and economic infrastructure to a region through schooling, health care, university research funding, and partnerships with idea-generating and idea-using groups. This infrastructure provides a foundation for attracting people with ideas to contribute to a region, and increases "social capital" that is important to enterprise innovation (Cook and Wills 1999).

The debate about mercantilistic approach has had the unusual effect of uniting both the "left" (neo-mercantilistic) and the "right" (laissez-faire) ends of the political spectrum, but for different reasons. The left argues that tax money is being spent on science and innovation for corporate profit that will not benefit the region, while the right argues that government involvement decreases the mobility and exchange of information and employment opportunities. As a result, both perspectives question the subsidization of university education and research, which produces talented individuals with ideas, who are increasingly mobile and willing to move to other regions (Reich 1992, 2001). At the same time, both the left and the right feel the specific region (local, provincial, or national) is losing in the global economic game of value-added production. From these arguments, other issues emerge such as the "brain drain" of high technology and scientific talent to low taxation and high income regions such as the U.S.; multinational corporations bargaining with what appear to be increasingly powerless local governments; charging students full cost for education; and increasing anxiety and protest concerning world trade organizational arrangements that drain local economic and social resources. These issues raise concerns about how knowledge and knowledge workers can be produced and employed within our geographical regions along with the commercialization, exploitation, and sale of value-added technology and the ideas embodied within.

These perspectives, particularly the neo-mercantilist one, incorporate assumptions about the relationship between ideas and economic development. It is to this end that I ask two particular questions: (1) What is the relationship between scientific and innovation/technology ideas (globally and regionally)? and, (2) What can public policy makers do to support the development and use of regional and global ideas in local economic development? The following related questions follow: Does government-funded university research and education increase regional economic activity through idea creation, sharing and dissemination? What is the relationship between local and global science during scientific discovery? What is the relationship between local and global technological development and its effect on regional economic development?

Confronting the dilemmas inherent in these questions and their answers will have important implications for issues such as government fiscal and monetary policy, funding in education, the likelihood of controlling success or failure in economic development through science and innovation policy, public and private financing of basic and applied research (Muniagurria 1995), the funding of technological development, the protection (and sharing) of technological know-how, the role of science and innovation in product/service innovation, and the effect of these activities on regional/national/international economic development. I use the term "dilemma" to describe the tension inherent in the effects of public policy on idea development and use, and regional economic development:

- knowledge cannot be objectified and protected within a single region;
- scientists do have international allegiances to the pursuit of knowledge;
- "symbolic analysts" who work in scientific and technological enterprises are very mobile between regions (Reich 1992, 2001);
- multinationals investing in a region do have diffuse ownership, and therefore allegiances across different regions;
- many products, services and organizational forms rely on a complex linkage of knowledge and coordination across organizations and regions (Galbraith 1986; Robertson and Langlois 1995).

As specific examples, I focus on the role of public policy in the electronics industry in Taiwan (Romer, Basu, Selowsky, and Srinivasan 1992), and in the canola-based biotechnology cluster in Saskatchewan (Phillips, Chapter 2).

My main conclusion is that a balanced perspective, what I call "co-operative mercantilism", needs to be fostered by public policy makers to effectively handle the emergent and organic sharing of ideas among science, innovation and business enterprise, and to counteract the mercantilism of multinational organizations (Phillips 1992). To do this, governments should increase the regional and local value of a core region's products and services by fostering an environment that strategically links regional scientists and entrepreneurs with their global counterparts. Governments must be willing to allow regional ideas to be shared and coupled with other international groups in order to ensure that these ideas draw on and add to the vast array of human knowledge required to produce advanced products and services (Rotenburg

2001). Quoting Phillips on local innovation clusters, "they must be nurtured with an appreciation for their partial and incomplete nature. Fundamentally, they are part of a global innovation system, and cannot thrive if cut off from lifeblood of the system – ideas, skilled labour and collaborative platforms" (Chapter 2 of this book: page 56). Key to this collaborative exchange is the university, with its access to researchers who are connected and associated with the international community, and who have the time to pursue knowledge sharing in an environment free from the day-to-day pressures of industrial enterprise.

At the same time, governments must have some power to direct the private flow of ideas between foreign corporations and local industries. The power to dictate explicit arrangements with these companies comes from their interest in accessing resources (human, material, or other ideas) or markets in the region. The spread of a particularly powerful idea into local production has been shown to provide almost 100% benefit to the region, with little cost to the foreign company. To be effective, however, local circumstances dictate that missing industries that are necessary to support other industries can most benefit from particular arrangements, and public policy makers need relative freedom from corporate and international influence to make these decisions. In addition, governments must be involved in maintaining the social infrastructure (schooling, university research, health care, and transportation) that supports the long-term idea generation and the productive capacity of a region. Implications for public policy and innovation, entrepreneurship, and economic development are discussed next.

Terminology

At the outset, I would like to clarify some of the terms used in this paper in order to avoid possible conceptual confusion in the discussions related to ideas and economic development.

I use the term "science" to refer to more theoretical, as opposed to applied ideas, often generated more for understanding a phenomenon than for direct application in commercial or social realms. In the past, applied ideas were not necessarily derived from university-based basic research, although the strong connection between innovation and basic science has changed recently. The term "innovation," as opposed to "invention" and "technology," refers to the infusion of new ideas into productive use and refers as much to the social conduct of the entrepreneur in the innovative process as to its physical embodiment in its "invention" or "technology" (Dassbach 2000; Amos 1996; Rogers 1995). This reference to the social context of innovation has become increasingly important in recent innovation research, as researchers have begun to realize the importance of social networks, communication channels, social systems, and human agency in the innovation process (Rogers 1995, 15).

> Past diffusion research has generally investigated each innovation as if it were independent from other innovations. This is a dubious assumption, in that an adopter's experience with one innovation obviously influences that individual's perception of the next innovation to diffuse through the individual's system. In reality, a set of innovations diffusing at about the same time in a system are interdependent. It is much simpler for diffusion scholars to investigate the spread of each innovation as an independent event, but this is a distortion of reality.

The terms "invention" and "technology" refer more to the physical embodiment and the organizational (i.e., "soft") processes that allow for more efficient and effective production of goods and services. In a "hard" embodiment in an invention, ideas can be ported to other domains more easily, but still require soft knowledge to manage their implementation and application.

The term "ideas" refers to all of the preceding terms: science, technology, innovation, and invention. Each area carries with it certain theoretical and applied advice about the social and physical world. "Entrepreneurship" is defined as those activities that cement science and technology together into an innovation that transforms organizational production.

The term "intermediaries" refers to all the various individuals, factors, and processes involved in the link between science and regional economic development. These include technology, innovation, entrepreneurship, input products/services, knowledge, and human expertise.

Finally, the term "regional economic development" refers to the development of human resources, knowledge, technology, products, and services that enhance the economic and social well-being of those individuals in a particular region, through the consumption, application, and export of these resources to other regions. It is the relationship between ideas and regional economic development, and its relationship to the above concepts, that I turn to in the following section.

Ideas and Economic Development

The relationship between ideas and economic development has grown closer and more complex as modern technology has become increasingly reliant on theoretical science (Latour 1987; Laidler 1998). This complexity continues to challenge public policies that try to foster, explicitly develop, and manage science and technology for regional economic and social advantage. In the following section, I attempt to roughly outline how scientific and innovative ideas affect the productive capacity and cost structure of an organization within and across regional boundaries. The analysis lays the groundwork for discussion of specific public policy cases and of the implications for public policy in the following sections.

With the numerous possible applications of theoretical science, combined with the increasing efficiencies in organizational production, we see a greater complexity in the relationship between these two important areas of idea generation and application. Starting with the traditional economic focus, Figure 1 provides a general model of the firm, exploring the relationship between human and material resources, science, technology, business organizations, product/service development, and the economy across local and global markets. The figure provides a general model of a single organization, operating either within a specific region or in multiple regions. The organization consumes three products/services and produces four products/services for either firm or consumer consumption. Using general economic concepts, each internal and external transaction involves human resources with particular prices (wages) for inputs. In turn, each demands a certain price (wage) by the organization producing these items.

After paying for the inputs, and receiving compensation for its outputs, organizations must make a reasonable profit or, in the case of not-for-profit organizations, break even. As a result, organizations themselves locate where needed resources are relatively inexpensive and/or plentiful, and near markets that are relatively cheap and/or plentiful to serve. Important considerations in many of their decisions are transportation and logistical costs. At the turn of the 20th century, organizations grew in size in order to increase economies of scale by taking advantage of the one great administrative innovation of its time: specialization.

In comparison with organizational production of that time, specialized knowledge has become paramount in efficient and effective production, thus increasing the specialization, division of labor, and coordination required (Galbraith 1986; Laidler 1998). This specialization has had important implications for production. First, most organizations have become increasingly specialized cogs in a larger and longer supply-chain – from raw materials to final product/service. A very small percentage of the market now resembles the independent capitalist who produced a single commodity product, as so often discussed in classical economics (Galbraith, 1986). Second, human expertise in both scientific and technological development has become the most important cost to organizations, whether in the form of consulting, the purchase of specialized materials that involve extensive human knowledge, or in the purchase and use of science and technological processes.

Extending our model to explore an organization from a regional perspective, Figure 2 examines the organization across regional boundaries.

Figure 2 paints a detailed and highly networked picture of the business organization in a knowledge-based economy, operating across regional boundaries. In this diagram, we see that the firm acts in the region's best interest if it can minimize the organization's import costs and maximize the price of export revenue from the selling of material, consulting services, science and innovation, and consumer products to non-regional markets. Profits from such ventures can then be reinvested locally for the recruitment, purchase, and development of new inputs.

Interesting implications emerge from Figure 2 that challenge the extreme positions of the neo-mercantilist and the laissez-faire approaches. First, local ownership is unnecessary for regional economic advantage. In this model, dividends and profits to local shareholders are only one part of the economic pie. In many instances, it may be more regionally advantageous to have a company, whether locally owned or not, to provide and consume ideas and human resources in its production of goods and services (Romer, Basu, Selowsky, and Srinivasan 1992). In this way, foreign organizations may also be a conduit to global ideas and capital at little (or possibly no) cost to a region. They may also be willing to spend the time required to infuse these ideas into regional products and services, generating local expertise, products, and services for local and global consumption. In contrast, the total outsourcing of production by a local company to other regions would mean that the local company's innovation, investment and production may never be felt in the region.

Second, innovations that can draw on material and human resources that are already available in a region may produce considerable benefit (de Gouvea and Kassicieh 2001). In Alberta's case, the expertise and technological know-how required in raw material extraction and processing is a considerable source of economic advantage, and has

Figure 1

Organizational inputs and outputs abstracted into four categories: consulting, material, science and technology, and consumer products.

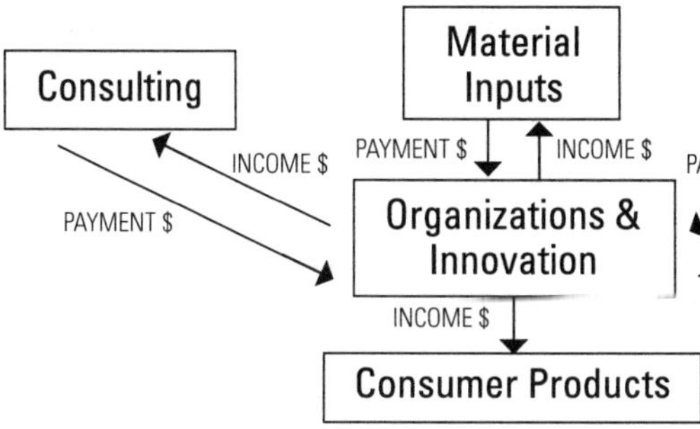

Figure 2

The model of the regional firm with inputs and outputs divided into local and global categories.

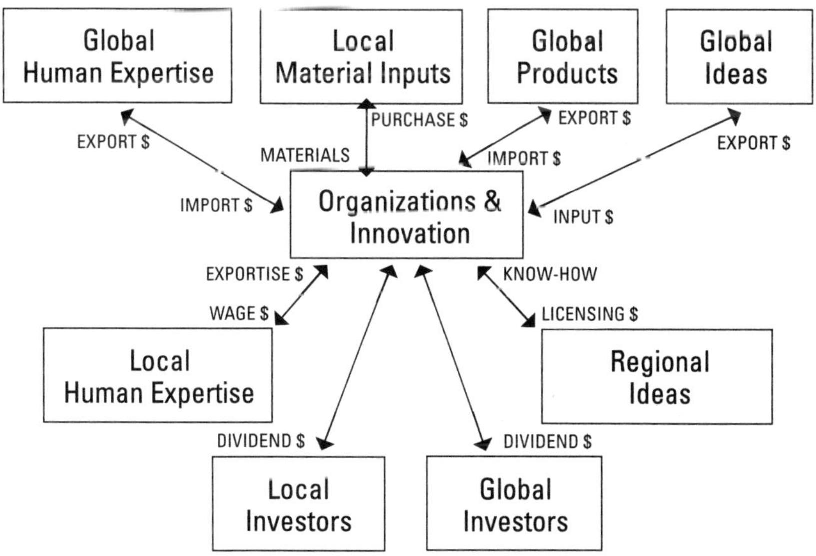

formed the basis of Alberta's advanced economy, which has been the spark of most of its economic growth (Reich 1992). Those organizations that can use and export the technology of resource extraction in Alberta, for example to other oil and gas regions or other industries within Alberta, will be the most beneficial to the region. The question remains: How portable is oil and gas knowledge across regions and alternative industries?

Finally, important in Figure 2 is the definition of a region. Is the region local, corporate, municipal, provincial, territorial, national, or international? If a region is defined globally, then Figure 2 becomes Figure 1. In addition, since a region is usually defined by space, it masks the unequal distribution of the benefits of organizational production and economic activity to different members within a region, and diverts attention away from disparities in outcomes within a region (Reich 1992). The definition of the region is also an important issue in scientific work, where the community of idea generators is often international. To explore the science/technology relationship in more detail, Figure 3 examines theoretical and applied ideas across science, technology, organization, production, and economic development.

Of particular interest is the relationship between science, technology, and innovation. A number of factors have increased the complexity of the science/technology/innovation relationship. First, in the past, the development of simple inventions, generated independently of scientific ideas, sometimes produced unexpected surprises in their final form and use. Historically, the science of technology was relatively primitive, allowing technological innovations to be created in the absence of basic science. Thus, a single inventor could often understand the science and the technology. These so-called "empirical inventions" had little or no foundation in theoretical science. But even with a simpler and sometimes nonexistent relationship with science, many inventors were unable to predict the eventual applications (i.e., innovation) of their technology. Long time horizons from scientific discovery to societal application seriously increase the complexity of the final technological form (Basalla 1988; Petroski 1994). Thus, Faraday's basic research on electricity founded numerous innovations in the electricity industry. However, at the time Faraday did not know or speculate about its future application (Laidler 1998). As a result, the diffusion and appropriation of technology by various groups, beyond the original inventor, is complex and emergent (Orlikowski 1992). This is still the case today despite our knowledge of this emergent relationship.

In addition to emergence in technological application, there are two factors that have increased the complexity of a relationship between science and innovation in contemporary times. First, today's technology rests on a deep and complex relationship between branches of the science and various social networks that support innovation (Galbraith 1986; Latour 1987; Rogers 1995). No one person can be conversant in

Figure 3

The relationships between science and technology within and across regional boundaries.

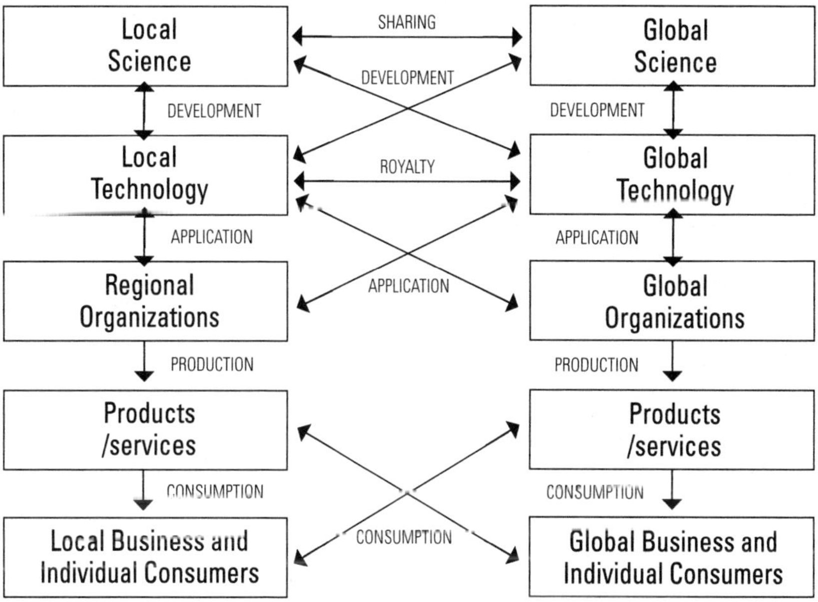

all of the scientific and technological ideas involved in its successful application in production. Each person has specialized so as to be able to contribute to a small fraction of a rapidly advancing field. Second, the politics of innovation indicate that shifting power alliances and beliefs play an important and unpredictable part in the motivation behind ideas and their application (Latour 1987, 1994; Noble 1999; Thomas 1994).

Public Policy in Action

Given the previous analysis of the relationship between ideas and productive capacity, I would like to draw your attention to two particular cases of public policy action in the management of ideas in regional economic development. Both demonstrate a cooperative mercantilistic approach – cooperating and negotiating explicitly with foreign groups to encourage regional economic advantage.

The first arises from a discussion by Romer, Basu, Selowsky, and Srinivasan (1992) about Taiwan's electronics industry and its rise from third world conditions to an idea generator and exporter. The second arises from a discussion by Phillips (Chapter 2) of the biotechnology industry in Saskatchewan.

Taiwan's strategy, began in the 1950s as a strategy of import substitution through traditional protectionist measures to foster local industry. However, instead of isolating industries from global networks, the government actively sought out contracts with foreign companies to transfer applied ideas into Taiwan and stimulate competitive manufacturing. Starting with a manufacturing base in radio assembly that was protected by restrictions on imports, they developed a technology agreement with a foreign firm. In addition, an export processing zone (EPZ) was set up to allow companies unrestricted, tariff-free imports of machinery and materials in return for providing no restrictions on ownership or repatriation of profits. Possibly fueled by increasing tensions between the U.S. and China, and American desire to have a free trading partner in the region, 24 U.S. firms had set up operations in Taiwan by 1965.

To encourage free discussion between industry competitors and researchers, the Taiwanese government formed electronics working groups that assisted in marketing, training personnel, and product expositions. An industrial technology research institute, and an electronics research service organization (ERSO), supplied the first Basic Input Output System (BIOS) used in Taiwanese computer clones.

Because of its large market and its experienced and inexpensive labor force, the government was still able to enact restrictions designed to protect and develop domestic industry. Proposals for foreign investment outside an EPZ required approval by government, and were subject to conditions that limited sales in domestic markets. For example, the government forced Singer Sewing Machines to purchase 80% of its parts from Taiwanese industries within a year of entering the market. This facilitated technology transfer and helped to upgrade the domestic components industry. When Proctor and Gamble opened a plant in the 1970s, they were required to export 50% of their output. Thus, the government provided free trade in ideas and their application, but directed the trade of finished goods and products. As a result, foreign ideas helped the Taiwanese economy in three ways: by creating a surplus for unskilled labor, by producing and using specialized human capital through on-the-job training, and by increasing the productivity of research. In this sense, Taiwan was able to capture close to 100% of the gains from the idea's application in the region. The process proved to be far more directly effective than increased funding of schooling which, if cut off from the world's applied knowledge, would have been ineffective.

Moving to the biotechnology cluster in Saskatchewan (Phillips, Chapter 2), the industry was founded on an established base in agriculture. Phillips uses the classical trade entrepot to describe the intricate interdependencies between Saskatchewan and global ideas. As exemplified by Hong Kong, the entrepot imports most of its inputs, adds value to a product or service, then exports the products as intermediate goods for further production. The Hong Kong approach is very similar to the Saskatchewan case. The Canadian record as a lead innovator and early adopter of new strains was very good, but many of the development in new canola-based varieties and applications research were happening elsewhere. The trade entrepot method allowed the Saskatchewan biotechnology industry to draw on the world's expertise to increase the value-added potential of the region's value-added research.

Most Saskatchewan biotechnology firms have developed extensive networks with collaborators and competitors. Much of the networking happens through deliberate public funding of universities, which offer a non-competitive environment for sharing ideas and the National Research Council (NRC), which is able to be the central learner from all groups. Most of the expert labor force is trained in Canada, but some of the more highly skilled employees need to be recruited from elsewhere. Particularly important in the innovative process are foreign and Canadian research stars who have been important in the generation of new start-ups. The public system has been the primary source of financing, and extensive networks have been developed to diffuse new strains quickly to local producers.

These two brief cases provide us with various public policy and industry responses to the science and regional economic linkage. They provide insights into the public policy implications and dilemmas that follow.

Implications and Dilemmas

Given the complex relationship between basic and applied ideas in regional economics, and the role of various intermediaries such as technology, entrepreneurship, and innovation, what public policy actions are possible? As has been argued, both the neo-mercantilist and the laissez-faire approaches to public policy have elements of effectiveness, but in their extreme forms are inappropriate in addressing the various links between ideas and regional economic development. Both poorly address the complex dilemmas of public policy action in managing the complex lattice of local and global idea generation and application. My words echo Phillips (Chapter 2) in saying "that no single firm or region can truly be viewed as self-sufficient or self-sustaining." At the same time, no region can expect that an abdication of public policy through a laissez-faire approach will automatically yield regional economic benefits.

However, the public policy dilemmas which attract people to these two extreme approaches remain. The following section highlights these dilemmas, along with possible recommendations for public policy that flow from cooperative mercantilism. Cooperative mercantilism pursues scientific, technological, and economic advantage in cooperation and negotiation with other regions' interests, realizing that soft self-interest avoids the pitfalls of trying to hoard and protect an idea whose relevance is quickly eroded by time and isolation.

Local vs. global ownership

The first dilemma can be summarized by juxtaposing the following two statements: (1) local ownership keeps ideas and production in the region, vs. (2) global organizations gives us access to the most advanced production processes and product development.

There is a widespread belief that local ownership is crucial to regional economic development. But we must be careful with regard to what is to be owned: is it raw resources, materials, labor, or basic and applied ideas? Although local ownership does provide revenue to local investors in terms of dividends and profits, there are many other benefits that foreign companies can provide to a region that are equally if not more important. These include: ideas, employment of skilled labor, the use of locally produced inputs, and the attraction of skilled foreign workers. Both the Taiwan and Saskatchewan cases suggest that careful introduction and contracting with corporations and world class researchers to use and further develop their ideas can be very beneficial to a region. However, a mutually beneficial relationship will not happen automatically and must be explicitly managed by public policy makers. Therefore, both the protectionist and laissez-faire approach to public policy are prone to failure in their extreme form.

Protecting vs. sharing ideas

The second dilemma can be summarized by juxtaposing the following two statements: (i) protectionism keeps ideas in our region for our benefit; vs. (ii) sharing and connecting these ideas with others worldwide is what produces their value; "no idea is an island."

The desire to protect knowledge has been important throughout history, perhaps as a result of warfare and the need to protect oneself from the enemy. Business and governments appear to have adopted this secretive and protective attitude toward ideas in an attempt to hoard and apply knowledge exclusively within a region's production, or for the high-priced sale of it to other regions.

Certainly, copyrighting, patenting, trademarking, and trade secrets represent individual and public policy attempts to ensure that ideas and their expressions are both protected and shared. These encourage inventors to invent and reap an adequate reward for their activities, but also to quickly share the ideas with others so that they may improve on and continue the development of those basic ideas. However, the protectionist response has been to treat ideas as a commodity to be hoarded for regional advantage at the expense of those both inside and outside the region who may benefit most from their application and further development. This is particularly true in comparing developing and developed nations, the latter trying to create global standards for intellectual property protection, the former avoiding any legislation on intellectual property (IP) and allowing widespread copying and use of developed nations' ideas without penalty to internal violators (Pasco 1998).

Regardless of one's stance on this issue, it is obvious that most of the world's current knowledge has relied heavily on the founding ideas and the thoughts of others. For the most part, until the patent protection of the late 19th century, many people created and contributed their ideas to the public domain, often with little regard for personal gain.

This trend is changing, both for the better and for the worse. In a worst-case scenario, we can imagine a world where every individual within and between regions decides to protect and keep secret its basic and applied ideas. The result would be scientific and technological paralysis. With increasingly aggressive corporate, university, and regional policies in this direction, this world of parochial regional protection could become a reality. However, it is an idea's coupling with other ideas that produces value. By itself, an idea withers. Both the Taiwanese and Saskatchewan cases demonstrate the role of government-funded university and research institutions in the breaking down of competitive secrecy and in the sharing of knowledge between competitors.

The basis of our modern production system is ever-increasing specialization that allows us to dig deeper into specific topics, yet requires cooperative coordination to bring these areas together in addressing complex production processes. As a result, both the neo-mercantilist and laissez-faire approaches to knowledge are detrimental to regional and global economic development. Especially troublesome with the neo-mercantilistic approach is the separation of regional ideas from national and even international research, given the specialization and internationalization of scientific knowledge and the need for technology and science to interact with each other in order to produce anything valuable. In addition, the neo-mercantilistic approach suffers from pragmatic and ethical shortcomings, involving, for example, the enforcement and protection of ideas, the economic harm caused by the lack of widespread adoption and use of ideas, and the growing disparities between richer and poorer regions because of unequal access to advanced ideas.

The laissez-faire approach, however, also poorly handles the matter of support of the skilled scientists and technology developers who have the most mobility to move across regional borders, and fails to ensure that a region benefits from the exchange of local and global knowledge. What is required is a policy that simultaneously explores cooperation with international groups for the sharing of knowledge, while supporting the development and exchange of key technology with the partnering regions to further economic growth and social development. Also, important in this strategy is some method of encouraging those who are most able to move across regions to take a greater stake in the development of a region's quality of life (Reich, 1992).

As a result of specialization and the need to work with global research knowledge, strategic linkages are extremely important in complex scientific and technological fields. Scientific and technological progress rely on access to the specialized knowledge of the global community. Government efforts should be aimed at increasing access to, and partnerships with, this global community. This scenario implies that regional economic development must not be a zero-sum game, except for those so far removed from the action that it is impossible for them to participate. Instead, it should be a cooperative developmental effort between researchers and corporations in many regions. It suggests that governments and organizations should work closely in involving international groups and organizations in strategic partnerships to harness worldwide knowledge, especially in those areas with low growth and high poverty (Baland and Francois 1996).

Related to the sharing of ideas is the issue of the unequal distribution of benefits to individuals within a region. This situation has raised more important issues in public policy (antitrust legislation, taxation, redistribution of wealth, and the regulation of large multinational organizations who are demanding ever more from regional governments and communities in exchange for relatively fewer benefits) than in regional strategic advantage. The question remains: How much of the benefits from ideas should accrue to the community, the corporation, and the inventor? This question becomes difficult as knowledge extends beyond the region and is combined globally, and as the profits and domination of a global market potentially occurs through chance as well as management action.

Public or private funding of high tech talent

The third dilemma can be summarized by juxtaposing the following two statements: (1) high tech research should be funded by those individuals and organizations who will benefit directly from the research; vs. (2) public funding is required to attract and generate the global and local networks for knowledge production and application.

Many provincial and national governments have taken a mixed approach to handling the issue of university funding. For example, the Alberta government has recently begun an initiative to recruit and fund top-notch researchers in the province through grants, suggesting that stimulating homegrown research by recruiting external researchers will bear more fruit than buying that research from abroad. At the same time, various university tuition fees are being raised to compensate for reduced government funding of university education, a trend which is based partially on increasing belief that a university education is a private good.

The underlying concern is the proximity of the region to highly educated people involved in the generation and application of new ideas. Do we need regional research stars who generate advanced theories in physics in our region to benefit from some form of first-mover advantage? Or can we import ideas freely across our regional boundaries, and employ them without penalty? Is homegrown scientific and technological development a necessary requirement for regional economic advantage? Although the proximity between scientific and technology developers may be an asset, some knowledge, particularly the basic knowledge, can be fairly mobile across regional boundaries.

Clouding this complex debate are cries about the brain drain of homegrown high tech talent siphoned from Canadian schools to the U.S. At the same time, some groups claim that we are buying expertise from individuals who may have the least allegiance to our region: those whom Reich (1992) calls the "symbolic analysts." Continual discussions of the brain drain that focus on income disparities between Canada and the U.S. along with the idea that more funding will cure this problem, point in one direction. The discussion about symbolic analysts – mobile groups who most often reflect a neo-conservative agenda – arguing for minimal government involvement and reduced barriers to the free flow of ideas and human ability, point the opposite way.

The obviously complex and mixed solution is to acknowledge that an educated work force with access to university researchers actively involved in world class research is important to economic growth and the attraction of high tech talent and business into a region (Acs 1996). Reich specifically mentions that these people, although they work in a global market, must live in a community. Others suggest that Canada's unique social and cultural infrastructure stems the outflow and may in fact produce an opposite inflow of knowledge from other parts of the world, including the U.S. (Kesselman 1999). Providing such an inflow with infrastructure support is a part of the strategy – and Canada is quite good in this area when compared to the U.S. Unfortunately, continual focus on the brain drain reinforces the emphasis on wages as the only object of interest.

A related issue is whether the proximity between the science and its application is important for economic benefit. Scientific and technological applications occur within a social network, and the more effectively a region can find and fund people connected to a global network of leading-edge ideas, the better off the region may be. But we must also realize that certain ideas can be bought, and applied through technology that requires little human involvement. New ideas and entrepreneurial applications of ideas require a strong social network, timing, and connection with a global community of scientists to be effective. Given the global specialization of this knowledge, government policy cannot feasibly encase an entire area of research within a region. It can, however, bring in a few key individuals and provide funding to forge connections between regional and global scientists and technology developers to maximize scientific and technological development. It is within such a rich virtual network that residual benefits can accrue to both the local and the partnering regions. And those who decide to leave can be seen more positively as contributing to the economic development of other regions, and as someday possibly returning with new ideas from abroad. Certainly, many Asian countries such as Singapore have managed this process more explicitly by requiring a return residency in the home country for those who have completed their training abroad.

Basic vs. applied science

The fourth dilemma can be summarized by juxtaposing the following two statements: (1) basic science has the power to fundamentally alter the basis of economic activity; vs. (2) only over the very long term, at great risk, and in coordination with science and knowledge across many regions, can science fundamentally alter the basis of economic activity.

There is considerable public tension with respect to funding basic as opposed to applied science, with some arguing that only research working directly toward practical application should be funded. Taiwan demonstrates the value of working only with applied knowledge in regional economic development. Advocates of this approach also argue that basic research is too easily copied and can be used by others for economic advantage. As a result, governments have been, and continue to be, wary of basic research that is removed from immediate application.

During the early stages of scientific and technological development, the results are unpredictable. There is little guarantee that an idea will lead to technological development and economic advantage, and that the results will be commercialized and used in a region. Also, basic researchers are committed to global scientific advance. They ask why, often with little interest in the how. Therefore, the knowledge produced is broad in its application, and due to the peer-reviewed publication process, may be hard to protect.

Scientists tend to work at the international level, and their allegiances are to the international community and market. By fostering and developing a scientist's international links, we can allow regional participants to interact with the global knowledge network, but their knowledge can also be applied locally to multinational or regional companies working in the area. However, we suffer a public policy problem with these symbolic analysts, since their allegiances are to international audiences as opposed to regional interests. Still, the positive externalities of working beside and with these people, and of being the first to uncover basic scientific discoveries and to potentially commercialize their various forms in technology, could benefit the region – as in the Saskatchewan biotechnology example. Basic research can result in long-term and residual benefits, which can alter the foundation of existing production methods. Corporate research, on the other hand, despite significant strides, will have a tendency to restrict basic science and the spread of innovation. Therefore, as we see in the Taiwanese case, corporate research prevents the free exchange of information and partnership. One solution may be for governments to fund research at different levels, depending on the local, national, and international outcomes of the research, with basic research having more long-term global impact and applied research having more immediate applications to local regions.

Related to the issue of basic vs. applied is the issue of soft vs. hard technology. The discussion of technology rarely focuses on the soft human expertise and skills required to tailor, implement, and operate it. Most often, ideas are softest in their most abstract and novel form within the scientific community, and later become hardened into know-how knowledge in later stages of development. Ironically, the hard technology, the physical form of the technology that most people associate with the term "technology," can often be sold as a package without requiring extensive human involvement in its maintenance and operation. Hence, hard technology may provide little economic value to the selling region over the long term, but potentially inexpensive and significant value to the regions that employ it.

As a result, it is the softer technology that may produce the most regional strategic advantage, by aligning itself closely with local human resources, and by employing regional consultants in the export of such technology. Potentially, hard technology combined with soft technology and consultative understanding will produce viable industries that will settle in a region, while the attraction of leading researchers to these softer idea areas may produce the kinds of start-ups described by Phillips in the Saskatchewan case. In addition, in those regions focusing on further social and technological adaptation, the appropriation and diffusion of ideas into newer technologies will be important in helping to manage and monitor this difficult and tricky relationship.

Radical vs. moderate diversification

The final dilemmas can be summarized by juxtaposing the following statements: (1) the basis of any modern industry is the infrastructure developed by previous and current industries; vs. (2) we need to develop a high-tech industry in pharmaceutical or computer technology to remain viable.

Certainly the Alberta oil and gas industry has spurred the advanced economy that now exists in the province. Reich (1992) argues that a solid base in manufacturing or natural resources is the important stepping stone for any advanced economy. The service sectors and advanced products provided by an economy, although important, are really only an outgrowth of this base. The question remains, however: Can and should Alberta moderately or radically diversify from this natural resource base to ensure stability despite the swings in oil and gas pricing? The answer rests largely on the viability of alternative industries, which in turn depends largely on an industry's connection with ideas, suppliers, and consumers of various products and services. Just as a species cannot survive unless it lives within a complex ecosystem that can support its functioning, the development of a radical or moderately new industry requires the correct introduction of various systems, some of them temporary, to create an ecosystem in which the industry can thrive. As a result, both radical and moderate diversification are possible, but radical diversification requires heavy and intensive public policy to sustain its early formation, while moderate diversification requires fewer resources. Both can lead to the production of advanced products and services that provide high value-added components.

For example, Alberta's focus on raw material extraction involves fairly sophisticated technology. Diversification strategies, therefore, should avoid trying to create entire scientific and technological structures that do not have the required historically developed infrastructures in place. Spin-off technologies are generated from technology developed by oil and gas companies, for example, by adapting products and service expertise to other markets. Therefore, the region could build on its success by migrating products and services that are already being sold to other potential markets. The same is true with science, in that scientific expertise that has already taken hold in a region is a good foundation for building a homegrown scientific community that feeds on both local and global expertise.

In the early 1980s, the Alberta Heritage Foundation for Medical Research embarked on a process of more radical diversification, through the funding of the biotechnology industry. Despite claims to be engendering in a totally new industry, it was in many ways building on an established medical research community that had been growing in Alberta throughout the century, and on the work at the University of Toronto in the early 1920s culminating in the discovery of insulin. In fact, the islet transplantation work for diabetes that was funded by the Alberta government had been occurring and struggling to obtain funding in Alberta throughout the 1970s. As a result, the Edmonton protocol in islet transplantation is internationally recognized as leading edge in the field of diabetes research, and has the potential to support a very substantial biotechnology industry in Alberta.

Conclusions

The relationship between ideas and economic development has been, and continues to be, increasingly complex. Despite this complexity, however, the tendency in public policy and discussion, has been to resolve these complex issues by adopting one of two approaches: the neo-mercantilist approach, which assumes that regions need to protect and exclusively use processes and tools that contribute to value-added products and services; or the laissez-faire approach, which asserts that regions should avoid intervening in the natural movement of ideas and people across regional boundaries. However, the tendency to hoard ideas inherent in the neo-mercantilistic approach is harmful to both regional and global interests in a world increasingly reliant on global knowledge from numerous individuals and groups. Ideas that are treated as commodities are eroded by time and by their isolation from other scientific, technological, and innovative concepts. The laissez-faire model, with its absence of public policy, ignores the negative and potentially exploitative influence of foreign and regional companies by failing to direct their activities in ways that are beneficial to the region. Such benefits include the distribution of profits to local shareholders, the transfer of ideas into the region at a reasonable price, the employment of skilled and semi-skilled labor within the region in productive activities, and the use of local expertise and material in production. Ideas do have value and cannot be treated as simply nebulous products that escape regional control. Their joint development and use with global partners need to be managed for regional economic advantage.

The specialization, coordination, and appropriation of the knowledge generated by science and its application is increasingly global. As a result, a number of complex issues face policy makers in addressing how to manage the generation, procurement, use and sharing of ideas to increase regional economic development. These dilemmas include: local vs. global ownership of ideas and factors of production, protecting vs.. sharing ideas, public vs. private funding of high-tech research, basic vs. applied research funding, and radical vs. moderate diversification of economic activity. Each issue poses particular dilemmas for public policy-makers. Each dilemma creates specific challenges for public policy that must be addressed in some form of balance between the opposite responses.

I conclude that instead of following a neo-mercantilistic or laissez-faire approach, governments must follow a cooperative mercantilist approach. They must continue to foster an environment of both basic and applied research that strategically links regional scientists and technology developers with their global counterparts in order to increase the value of regional ideas, products, and services. They must be willing to let regional knowledge freely couple with other global ideas, with an expectation that this will bring the vast array of global human knowledge to bear on value-added products and processes, and that it will also allow regional knowledge to form the basis for significant value-added scientific and technological discoveries.

I explored the cooperative mercantilistic policies evident in the Taiwanese electronics and in the Saskatchewan biotechnology industries. Both cases show clear evidence of public policy makers' attempts to open up the channels of communication with foreign ideas, while at the same time setting up conditions for the benefits to accrue to local and foreign groups. Both recognize the overwhelming number of ideas that are created outside the border, and the need to employ and build on global knowledge in their scientific and productive activities.

In addition to the specific conclusions concerning ideas and regional economic development, specific public policy implications are reached for entrepreneurs, intermediaries involved in idea generation and application who are important in producing regional economic development. First, entrepreneurship could be defined as a behavioral process by a group of individuals in creating or applying new ideas that are currently unused in the global context. Thus, scientists, public policy makers, and business people are all involved in entrepreneurial activity, in new idea procurement, generation and usage. In order to be entrepreneurial in a way that is beneficial to the region, these individuals must be exposed to new global ideas, otherwise, their entrepreneurial activity reiterates what may already have been created and applied elsewhere. Their idea generation sits in a barren vacuum and does not build on and use what is considered to be new within the global economy. Only through exposure to global ideas does their activity produce and use local resources in the most productive manner, or can novel (by regional and global standards) products and services be created by their efforts (either by cost or new products). New ideas and entrepreneurial applications of ideas require a strong social network, timing, and connection with the regional and global community of idea generators to be effective.

These ideas are fragile at this stage. But my purpose is to generate debate and discussion around the many possible forms of public policy intervention that could support the effective and equitable use of local and global ideas in regional economic development. I hope that I have raised enough interest and reaction to set this process in motion.

References

Acs, Z.J. (1996). Does research create jobs?. *Challenge* 39(1): 32.

Amos, Orley M. (1996). Chapter 3: The long wave cycle. (Web page) http://amos.bus.okstate.edu/pole/GPCCH03.html.

Baland, J.M. and Francois, (1996). Innovation, monopolies and the poverty trap. *Journal of Development Economics* 49(1): 151-52.

Basalla, G. (1988). *The Evolution of Technology*. Cambridge: Cambridge University Press.

Cook, P. and Wills, D. (1999). Small firms, social capital and the enhancement of business performance through innovation programmes. *Small Business Economics* 13(3): 219-34.

Dassbach, C.H.A. (June, 2000). Entrepreneurial motivation and the clustering of innovations: A reconsideration of Schumpeter's theory of the Kondratieff cycle. (Web page) http://www.1-888.com/longwave/oladcha2.html.

Galbraith, J.K. (1986). *The New Industrial State.* New York: Mentor Books.

de Gouvea, R., and Kassicieh, S.K. (2001). Resource-based innovation: The case for the amazon region. *Engineering Management Journal* 13(1): 28-32.

Hagelstam, J. (1991). Mercantilism still influences practical trade policy at the end of the twentieth century. *Journal of World Trade* 25(2): 95-106.

Kesselman, J.R. (1999). Policies to stem the brain drain – without Americanizing Canada. In *The Brain Drain: Causes, Consequences, and Policy Responses.* ed. F. Mihlar. (Web page) http://www.arts.ubc.ca/cresp/Brwebab.htm

Laidler, K.J. (1998). *To Light Such a Candle: Chapters in the History of Science and Technology.* Oxford: Oxford University Press.

Latour, B. (1987). *Science in Action: How to Follow Scientists and Engineers Through Society.* Cambridge: Harvard University Press.

Latour, B. (1994). Where are the missing masses? The sociology of a few mundane artifacts. In *Shaping Technology/Building Society: Studies in Sociotechnical Change.* ed. W.E. Bijker and J. Law, 225-58. Cambridge: MIT Press.

Muniagurria, M. E. (1995). Growth and research and development. *Journal of Economic Dynamics and Control* 19(1,2): 207-36.

Noble, D. (1999). *The Religion of Technology: The Divinity of Man and the Spirit of Invention.* Toronto: Penguin Books of Canada.

Orlikowski, W.J. (1992). The duality of technology: Rethinking the concept of technology in organizations. *Organization Science* 3: 398-427.

Pasco, B.C. (1998). Technology transfer in a Ricardian mode: China's technological "osmosis" in theory and practice. *Studies in Comparative International Development* 32(4): 79-100.

Petroski, H. (1994). *The Evolution of Useful Things*. New York: Vintage Books.

Phillips, P.W.B. (1992). Whether free or fair trade, Corporate mercantilism rules the day. *Challenge* 35(1): 57.

Phillips, P.W.B. (2002). Regional systems of innovation as modern R&D entrepots: The case of the Saskatoon biotechnology cluster. In *Innovation and Entrepreneurship in Western Canada: From Family Businesses toMultinationals*. ed. J.J. Chrisman, J.A.D. Holbrook, and J.H. Chua. Calgary: University of Calgary Press.

Pinch, S., and Henry, N. (1999). Paul Krugman's geographical economics, industrial clustering and the British motor sport industry. *Regional Studies* 33: 815-27.

Reich, R.B. (1992). *The Work of Nations*. New York: Vintage Books.

Reich, R.B. (2001). *The Future of Success*. New York: Knopf.

Robertson, P.L., and Langlois, R.N. (1995). Innovation, networks, and vertical integration. *Research Policy* 24: 543.

Rogers, E.M. (1995). *Diffusion of Innovations*. New York: Free Press.

Romer, P.M., Basu, K., Selowsky, M., and Srinivasan, T.N. (1992). Two strategies for economic development: Using ideas and producing ideas; comments; floor discussion. *The World Bank Research Observer*. Washington.

Rotenberg, M. (2001). Confidence and e-commerce. *Organisation for Economic Cooperation and Development. OECD Observer* 224: 53-54.

Thomas, R.J. (1994). *What Machines Can't Do: Politics and Technology in the Industrial Enterprise*. Berkeley: University of California Press.

II
Entrepreneurship and Innovation

Chapter 6

Developing Market Economies: The Aboriginal Case in Northwest British Columbia

Ronald K. Mitchell and Eric A. Morse

Abstract

In this paper, we summarize Transaction Cognition Theory and the key propositions that flow from it and apply to the society level of analysis, as the theory bears on economic development – specifically, the creation of high performance economies among the native peoples in northwest British Columbia. Based on the concepts and analysis presented in this paper, we conclude that two contingencies will influence the likely success of economic development initiatives in northwest B.C. First, the successful negotiation of on-reserve property rights; and second, the acquisition by economic decision makers (leaders, venturers, and even the general members of on- and off-reserve native society) of the levels of planning, promise, and competition entrepreneurial cognitions that are sufficient for their roles in economic development.

Introduction

Understanding how to navigate from hierarchical to market-based economies has been, and remains, an important question for scholars interested in the privatization and entrepreneurial transformation of command economies (Mitchell 1992). While working on this problem in the Chinese context, we had the privilege of receiving a delegation of native leaders from northwest British Columbia. They were struck by the parallels between the economic development problems experienced during the Chinese transition to a market-based economy, and those presently confronting native communities.

In mainland China, the assumptions of state-sponsored socialism resulted in collectivization of production, with a command economic structure set in place to centrally plan and manage it. In northwest B.C., these native leaders observed that the provisions of the federal Indian Act have, through the reserve system, also resulted in elements of collectivization. Through the band council system and the regulations of Indian and Northern Affairs Canada (INAC), a command economic structure in which the assumptions of central planning and management have a strong influence is also in place. Both situations urgently require guidelines for developing high performance economies.

The command to market issues that presently confront aboriginal populations on-reserve in northwest (and on occasion elsewhere in) B.C. are explored in this paper. Herein we will we summarize Transaction Cognition Theory[1] and the key propositions that flow from it and apply to the society level of analysis, as it bears upon economic development – the creation of high performance economies. Native focus groups that we have conducted in northwest B.C. consider a high performance economy – and by extension, a working definition of economic development in the First Nations community – to mean "the processes that lead to prosperity and cultural well-being." As we theoretically develop and justify three key propositions, we provide an example of how both positive and negative outcomes in native economic circumstances can be explained by Transaction Cognition Theory. This paper concludes with discussion and observations that arise from the analysis.

Transaction Cognition Entrepreneurship Theory

Transaction cognitions consist of specialized mental models or scripts (Arthur, 1994a; Mitchell, Smith, Seawright, and Morse 2000; Neisser 1967; Read 1987) that guide individuals' responses to three principal sources of market imperfection: bounded rationality (BR), opportunism

[1] Although related, Transaction Cognition Theory should not be confused with, and is distinct from, Transaction Cost Economic Theory. The principles of Transaction Cost Economic Theory are utilized in Transaction Cognition Theory to help to identify and rigorously derive the thinking patterns (cognitions) that enable transacting at various levels of analysis.

[2] The terms bounded rationality, opportunism, and specificity have particular definitions within transaction cost economics (Williamson, 1985) that are adopted and utilized within Transaction Cognition Theory as follows:

- Bounded Rationality: The condition that leads to behavior that is intendedly rational, but limitedly so.
- Opportunism: Self-interest seeking with guile.
- Specificity: Non-redeployability (once time has been expended in the creation of a work, it is impossible to redeploy that same productive time for a different work).

Table 1: Some Attributes of the Contracting Process
(Williamson 1985: 31)

Behavioral Assumption			
Bounded Rationality	Opportunism	Asset Specificity	Implied Contracting Process
0	+	+	Planning
+	0	+	Promise
+	+	0	Competition
+	+	+	Governance

0 = absence, + = presence

(O), and specificity (S) (Williamson 1985).[2] Williamson argues that the contracting processes in the transacting world include: (1) planning, (2) promise, (3) competition, and (4) governance/hierarchy, depending (respectively in each instance) on the presence/absence combination of the foregoing market attributes (BR, O, and S), as shown in Table 1.

This framework suggests at least three sets of attribute/process relationships: (1) between bounded rationality and planning, (2) between opportunism and promise, and (3) between specificity and competition. Interestingly, although these relationships are inherently bidirectional, Williamson utilizes only one direction in his analysis of hierarchies vs. markets. He suggests, for example, that the absence of bounded rationality in the presence of asset specificity and opportunism implies planning. However, he underutilizes the complementary idea that planning should also reduce bounded rationality in situations characterized by those same two conditions (Simon 1979) (because better or worse planning affects the level of transaction costs that arise from bounded rationality). The same conclusion follows for market imperfections created by opportunism and asset specificity. Opportunism is expected to be affected by promise processes (e.g,. trust creation (Barney and Hansen 1994) among stakeholders (Agle, Mitchell, and Sonnenfeld 1999; Mitchell, Agle, and Wood 1997)), and specificity by competition processes (e.g. the adoption of a low-cost generic strategy (Porter 1985)).

Thus, the cognitions that individuals possess about planning, promise, and competition are expected to impact transaction costs, and therefore the success of transacting, where:

- planning is defined as the mental models that assist in developing analytical structure to solve previously unstructured market problems;

- promise is defined as mental models that help in promoting trustworthiness in economic relationships with, for example, stakeholders (Agle, Mitchell, and Sonnenfeld 1999; Mitchell, Agle, and Wood 1997); and
- competition is defined as mental models that can create sustainable competitive advantage, and are expected to impact transaction costs, and therefore the success of transacting.

Transaction costs are the costs of running the economic system. They are to economic systems what friction is to physical systems (Arrow 1969; Williamson 1985). Entrepreneurial opportunity (Kirzner 1982) occurs when entrepreneurs utilize planning, promise and competition cognitions to enact transactions that would otherwise fail due to the transaction costs. Entrepreneurship may, in this respect, be conceptualized as an essentially cognitive process (Mitchell, Smith, Seawright, and Morse 2000).

By definition, a transaction occurs when an individual creates a work (some product or service) and then enters into an exchange relationship with other persons for the sale or acceptance of that work (Gardner 1993) as illustrated in Figure 1. Transaction cognitions are the mental models or scripts (Arthur 1994a; Read 1987) that are utilized in this process. Thus, where the objective of entrepreneurship is to discover and enact successful transactions (Kirzner 1982), the job of the entrepreneur is to use market imperfections to advantage. This reasoning produces the Transaction Cognition Theory definition of entrepreneurship.

Under this definition, entrepreneurship is: the use of transaction cognitions (mental models/scripts about planning, promise, and competition) to organize exchange relationships (among the individual, the work, and other persons) that utilize the sources of market imperfections (bounded rationality, opportunism, and specificity) to create value (Arthur 1994b; Csikszentmihalyi 1988; Gardner 1993; Mitchell 1999; Williamson 1985). This is illustrated in Figure 2. The linkage between this conceptualization of entrepreneurship and native economic development can then be seen through the application of Transaction Cognition Theory at the society level of analysis.

The literature suggests that multi-level constructs occur in theories that can be generalized across levels (Rousseau 1985). As such, critical uniformities are required. In the case of Transaction Cognition Theory, three multiple-level sets of constructs represent the individual (as creating entity), others, and the work at multiple levels. As well, three

Figure 1: The Elements of a Basic Transaction

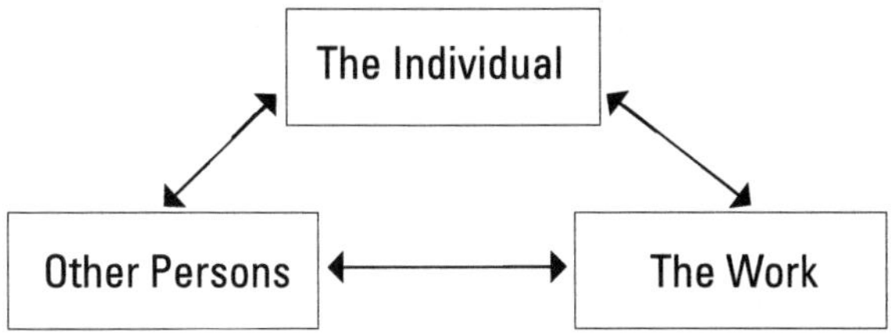

Based on Gardner (1993)

multiple-level sets of cognitions, planning, promise, and competition, suggest specific cognition constructs at corresponding levels.

As illustrated in Figure 3, at the society level of analysis, it is Gross Domestic Product (GDP) (as the creating entity) that, according to the values system of a society (the common attributes of other persons that shape the physical and behavioural artifacts of that society (Schein 1985)), produces the standard of living (Mitchell 2001). Based on prior work (Mitchell 1992), one set of planning, promise, and competition cognitions that operates at the society level of analysis includes productivity cognitions (planning), trust cognitions (promise), and value cognitions (competition).

Thus, the general transaction cognition model shown in Figure 2 can be specialized to represent these relationships at the society level of analysis as illustrated in Figure 3. In the next section, the justification for propositions that relate the elements shown in Figure 3 is presented. Cases from the native experience in northwest B.C. are also presented

Figure 2. The Transaction Cognition Theory Model at the Individual Level of Analysis

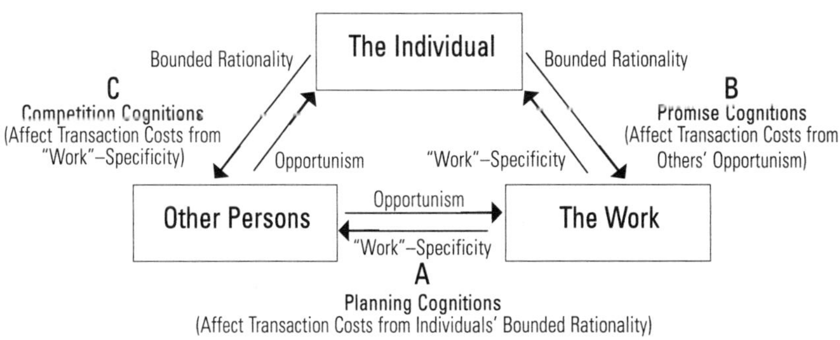

Based on Gardner (1993); Williamson (1985)

Figure 3: The Transaction Cognition Theory Model at the Society Level of Analysis

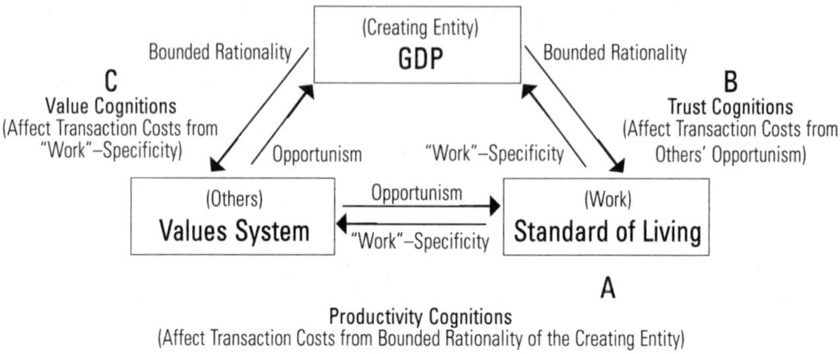

Based on Gardner (1993); Williamson (1985)

to illustrate the assertions made. The main objective of this analysis is to demonstrate how, at the society level of analysis, the development and fostering of specialized cognitions can, in fact, result in society-level economic development.

Propositions

With help from many individuals who live and/or work in northwest B.C. and are well-informed on native affairs due to their own aboriginal heritage and experiences or extensive work with the native community, several illustrations of the propositions that follow have been identified and are reported in this paper. These cases demonstrate how the theory explains both positive (+) and adverse (-) results of economic decision-making relative to the principles proposed.

Planning Cognitions

Transaction Cognition Theory suggests that economic frictions (transaction costs) are responsible for the grouping of transactions into the transaction bundles (firms, industries, economies, etc.) that result in the multiple levels of analysis previously discussed. Transaction cost economic theory suggests that the grouping process that creates organizational hierarchies is one of "discriminating alignment," and can occur "not only at the level of transactions but also at the level of nation-states" (Williamson 1996b, 332). Once such a transaction cost economizing grouping, or "hierarchy," exists, then an additional efficiency-creating process (coordinative alignment) occurs. This process aids the individuals who are bound together within that common transacting group (the individuals within a society) to economize upon transaction costs within that entity (Williamson 1991).

Unfortunately, it has been noted in earlier work that within the command economic systems studied to date, people's attention to the elimination of coordinative transaction costs is limited. As noted by one commentator in connection with the former Soviet Union,

> . . . the problem with Marx's work stems from his political beliefs and not his theoretical system. . . His political sympathies led him to focus on the macro-structures (in the economy) and largely to ignore micro-structures (Ritzer 1979, 35).

One danger of collectivization, then, is the "bureaucratization of economic life" (Lange 1938, 109), which leads to waste and inefficiency. For example, the management practices characteristic of the command economic system that the Soviets established and enforced in Hungary, resulted in waste and inefficiency characterized by ". . . excessive bargaining between supervisors and subordinates, pervasive distrust, the delegitimation of managers, and responsibility avoidance" (Pearce 1991, 75). Further, the fear of expropriation bred during an "era of confiscations" (Kornai 1986, 1705) created incentives for craftsmen, shopkeepers, and small business entrepreneurs to focus on "myopic profit maximization" (1986, 1706). As a result, a set of first order economizing costs (Williamson 1991) due to waste of all types operates within command economic systems to sabotage economic performance.

We can therefore expect to see, as a part of economic life under command economic assumptions, a set of transactional practices that increases the costs of running the economic system *within hierarchies* and exhibits no related benefit, either within the hierarchy, or at the societal level. Thus, when considering what might be done to facilitate economic development through the enhancement of market economies in northwest B.C. native society, attention to first order economizing of a coordinative nature is essential.

An additional implication of coordinative economizing is that *reversals* of the fundamental transformation (Williamson 1985) are also possible. Thus, assets maladaptively internalized (wasted in employment under hierarchy) might be returned to market governance. The possibility of fundamental transformation reversal suggests that tuning or adjustment type activities might be possible for economic agents which, through the elimination of waste, endow society with the redeployment benefit of underutilized assets, and reduce the likelihood that thus unburdened firms will fail. Hence, in the aboriginal case in northwest B.C., the following is proposed:

Planning Cognitions Proposition:
Economic development is associated with the planning cognitions that foster productivity.

Thus, it is expected that, where action is taken by economic actors to eliminate waste or inefficiency within existing economic structures (e.g., firms, society), privatization (the move from hierarchy to market) should be stimulated through reversals of the fundamental transformation while the prospect of firm failure is attenuated and material well-being is enhanced. Where attention is focused instead on bargaining over the allocation of resources within a command economic system, a lower GDP is to be expected, resulting in a lower standard of living. Both positive and negative illustrations of this proposition follow.

(+) The case of Northern Native Broadcasting Corporation (NNBC). The case of NNBC illustrates the veracity of the aforementioned planning principle. Several years ago there was a change in management at NNBC. Prior to the change, NNBC was operated much like a government department (depending entirely on a yearly subsidy, and operating with few planning cognitions: few, if any, financial targets and little accountability). Under the new management, waste and inefficiency were steadily reduced as effective planning, supported by a coordinated human resource approach, produced financial and market share targets that have been met or exceeded. As a result, the broadcast quality and innovation at NNBC has risen; the advertising revenues have increased dramatically; and the viability of NNBC as a fully capable market participant has been enhanced. As predicted by theory, these reductions in waste and inefficiency have stimulated the potential for privatization (the move from hierarchy to market), while the prospect of firm failure has been attenuated, and the material well-being of the organization and its stakeholders has been enhanced.

(–) The case of Nisga'a halibut licenses. The case of a halibut fishing license owned by the Nisga'a Nation Corporation illustrates the deleterious effects of command economic assumptions, and the resulting distrust, delegitimization of community members, and responsibility avoidance that ensue in an oversocialized command economic system. In this case, an entity controlled by four band councils (the Nisga'a Nation Corporation) held a valuable (five-figure value) halibut fishing license that it decided to put up for sale. When a member of the Nisga'a

nation came forward with an offer to buy, it created an internal debate. One key sentiment, which might be expressed as: "we don't want one Nisga'a individual to get ahead of another," resulted in the matter being shelved for at least a year. According to the individual recounting this incident, the situation created a lot of animosity within the community. Eventually, the license was sold through a broker to a non-Nisga'a native corporation. This sale incurred the additional expense of a broker's fee, and a loss in potential prosperity: the revenue-generating and culture-preserving capacity of the community.

According to the respondent, "there are plenty of individuals who have the potential to create jobs; but on-reserve there is a crab syndrome where others don't want to see their fellow members of society get ahead" (Anonymous interview April, 2000). The preference for tolerating waste and inefficiency within a society at the expense of the development of individuals' economic independence demonstrates a planning disability within the reserve system. In such circumstances, it appears that political considerations supersede economic ones. In this case, the political considerations fostered waste and lost opportunity for a member of the society. This suggests the decoupling of political and economic micro decision making within a market, since allowing transaction-by-transaction decision making to fall within the political arena (the command planning model) results in transaction costs from waste and inefficiency.

Where players in the economic arena use political means to influence the rules of the game, i.e., influence governments to create and sustain market imperfections that give profit to these powerful players *whether or not* they add the value of discovery to the system (Etzioni 1988), all members of that society lose. Outside the reserve system, governments that regulate market system economies act to preserve system integrity by removing the unproductive market imperfections that restrain trade. Actions such as the enactment and enforcement of antitrust laws are examples of approaches that are geared to leveling the playing field.

Therefore, to attain the benefits envisioned by adherence to the Planning Cognitions Proposition, Transaction Cognition Theory suggests that it is necessary for the plans within the governance of native society to abjure the involvement of political entities (such as band councils) in running businesses. The job of government should be to keep the game fair and to resist the temptation to misuse governance authority to obtain a share of game winnings, rather than to play within the game.

Promise Cognitions

Transaction Cognition Theory also suggests that promise cognitions are necessary for economic development a society. This is because one of the market imperfection-creating attributes of humanity is opportunism: self-interest seeking with guile (Williamson, 1985: 30). Opportunism fosters cognitions that produce social friction, which increases transaction costs due to moral hazard and distrust. Cognitions that reduce social friction through the promotion of trust can therefore be very helpful in creating a society with high levels of economic development, because they increase the promise that the expected benefits will be delivered.

Fundamental to a discussion of the freedom to make and keep economic promises, is the existence of property rights. John Adams, one of the framers of the U.S. Constitution, is reputed to have asserted that "property must be secured, or liberty cannot exist." Property rights represent commonly agreed upon standards that convey the rights of ownership. When they are clear and well defined, trust in transacting is enhanced.

Transaction cost economics usually assumes that property rights in a society are well defined and easy to enforce (Williamson 1975; Williamson 1985). However, for native individuals who live on-reserve, individual land ownership, a key property right, does not exist. Under section 20 of the Indian Act, a certificate of possession (CP), the closest thing to a right to own real property, may only be granted with the consent of: (1) the Minister of Indian and Northern Affairs (INAC) representing the Queen as the holder of title, and (2) the band council governing the reserve in question. This provision directly impairs the capability of individuals in on-reserve society to enter into credible economic promise relationships.

This missing capability to promise has pervasive implications for economic development. Lack of the right to own land on-reserve leads to a situation where the transactions that depend on the right to own real property (such as the financing of business premises or individual homes) fail due to definition and enforcement problems. Therefore, it seems useful to consider the application of the transaction cognition model to circumstances where property rights are neither well defined nor supported by a societal tradition of credible commitment to their enforcement (Williamson 1991).

Transaction cost economics suggests that under a weak property rights regime the fundamental transformation of market to hierarchy is induced at a lower level of asset specificity (i.e., markets fail), since inducements exist for transactions to be integrated (forward, backward, laterally) to mitigate expropriation hazards (Teece 1986). Alternatively, where governance structures are not readily alterable to safeguard transacting, it should be expected that "farsighted agents . . . recognize that their market development efforts will be expropriated . . . unless they are able to develop ties . . . which preclude the [expropriation] scenario from materializing" (Williamson 1991, 84). In essence, transacting agents must take a hostage (invoke a force-based promise) to raise the level of asset specificity for which a safeguard is mandated. It should be emphasized that the cost of such safeguards is an increase in transaction costs; the friction in running the economy goes *up*, contributing to dislocations, chaos, and other such drains on the societal well being that hinder economic development. In the aboriginal case in northwest B.C., such increased friction implies the following proposition:

Promise Cognitions Proposition:
Economic development is associated with the promise cognitions that foster trust by strengthening property rights.

Thus, to the extent that property rights are strengthened, opportunism-based transaction costs such as economic hostage-taking, the dislocation of productive economic resources, and the chaos within society are expected to decrease. To the extent that on-reserve property rights remain weak, political processes are expected to dominate in place of economic processes (Poelzer 1998), because of the need to reconcile uncertainties about control before trust can be expected to flourish. In the following cases, the efficacy of this proposition is demonstrated.

(+) The Sechelt Band negotiates its own real property rights. As noted above, in Canada, Her Majesty the Queen, represented by the supervising federal cabinet minister in the government party, has title to all reserve lands. Under section 91(24) of the Indian Act, federal jurisdiction is intended to preserve Indian reserves intact. Over the years, however, the Sechelt Band in British Columbia, being led by particularly farsighted individuals, has moved out of the traditional reserve system into a system of communal fee simple title for its former reserve lands.

In 1977, Sechelt completed the delegation of all available powers under the Indian Act through negotiations with the federal government. This gave it the right to exercise all the ministerial powers with respect to the lands on its 33 reserves. It went on to secure full self-government in 1986 as a result of Canada's Sechelt Indian Band Self-Government

Act. Thus, when, for example, prior to self-government, INAC refused (under section 53(1) of the Indian Act) to allow Sechelt to give an option on 300 acres of gravel land, the band gave the option anyway and, after self-government was affirmed, was able to honor the lease. Thus, Sechelt property rights under self-government (even though held communally and not individually) facilitated economic development.

(–) The case of building on-reserve. As noted above, by virtue of the provisions of section 20 no CP can be obtained, except by permission of the minister and the band council. Once a CP has been obtained, the courts have held that it may be transferred[3] with the approval of the minister as a part of section 91(24) (federal jurisdiction) lands. However, because of the lack of fee simple title, the band council – being susceptible to the political influence that inures in this governance process – can still increase the transaction costs, and cause transactions to fail. In several instances described to us by respondents, band councils have invoked mechanisms such as zoning, to frustrate the intentions of a band member in the transfer of a valid CP.

Further, according to the experience of one of the persons we interviewed during our field work, native individuals whose economic earning power easily qualified them for home mortgages when buying a house off-reserve could not qualify for a home loan to build a house on-reserve because the bank was unable to collateralize the loan with the home built due to the lack of fee simple title.

In these situations, the lack of clear and clearly enforceable property rights severely compromises transacting both on- and off-reserve. As with the case of planning cognitions, and as predicted, on-reserve promise cognitions tend to be highly developed, but are utilized primarily for political as opposed to economic purposes.

Competition Cognitions

Competition cognitions comprise the third set of mental models specified by Transaction Cognition Theory essential to economic development. As defined previously, competition cognitions are the mental models that create sustainable competitive advantage in market-based transacting. Competition cognitions are expected to impact the efficiency of transacting, and therefore the level of economic development, through their ability to help individuals within a society to align expectations about the work produced with the most effective market mechanisms.

[3] 1984 decision by the Supreme Court of Canada, opinion by Mr. Justice Pratt.

As also previously mentioned, transaction costs are considered to be the "costs of running the economic system" (Arrow 1969, 48), and may be viewed as "the economic equivalent of friction in physical systems" (Williamson 1985, 19). Transaction costs hinge particularly on the level of specificity related to the work component of a transaction. These friction/transaction costs are minimized by autonomous adaptations: the adjustments in human transacting procedures effected by strategic choices that influence the operation of the price mechanism and are made automatically as individuals transact with others in a market economy. Transaction cost economists suggest that high specificity implies high transaction costs, thereby implying a hierarchy (Williamson 1975). Correspondingly, low asset specificity implies low transaction costs, thereby implying market governance.

Where, due to the political reality (such as the existence and enforcement of the Indian Act), command-type decision-making is utilized (e.g., on-reserve governance), two types of maladaptations (conditions that result in higher than necessary specificity and therefore transaction costs) can occur as a result of the impediments to the operation of the price mechanism that command decisions cause. These errors prevent the efficient governance of transactions, as illustrated in Figure 4.

Where low specificity, as an attribute of transactions, implies market governance due to autonomous economizing, attempts to govern such transactions within a hierarchy (e.g., by exercising INAC, or band council control or influence) constitute the Type I, or "collectivization" error illustrated in Figure 4. To compare this to the case of the command economy of the former Soviet Union, the Type I collectivization error appears primarily to be the kind of error committed during the era of state-sponsored socialism, causing failures in large centralized units of production (e.g., agricultural communes that failed to feed the U.S.S.R. on the richest farmland in the world). In the case of the on-reserve economy, we can expect to see the Type I error in situations where business ventures have insufficient market support but are still attempted on-reserve.

Conversely, where a high level of site, resource, or physical specificity would normally, under autonomous economizing, lead to hierarchy as the most efficient form of governance, attempts to govern transactions related to these assets via market mechanisms (i.e., forced marketization) are also predicted to be problematic. In this instance, a Type II, or "privatization" error could also be made. The key to efficiency, then, appears to be the unencumbered autonomous operation of the market

Figure 4: Predicted Results of Interventions on Market Alignment, as Influenced by Level of Specificity

		Level of Specificity	
		LOW SPECIFICITY	HIGH SPECIFICITY
Command Economy Intervention	HIERARCHY ATTEMPTED	1 Type I Collectivization Error (hierarchy attempted where a market should be)	3 Efficiency: (Hierarchy alignment: autonomous economizing)
	MARKET ATTEMPTED	2 Efficiency: (Market alignment: autonomous economizing)	4 Type II Privatization Error (market attempted where a hierarchy should be)

mechanism. According to Transaction Cognition Theory, the key to the unencumbered autonomous operation of the market mechanism is that the market actors possess adequate competition cognitions; otherwise, maladaptations occur. Accordingly, in the aboriginal case in northwest B.C., the following proposition appears likely:

Competition Cognitions Proposition:

Economic development is associated with the competition cognitions that foster value creation through market alignment.

Thus, to the extent that competition cognitions enable a smoothly functioning marketplace, low specificity transactions should be managed with minimum friction, thus avoiding "Type I collectivization errors." And, to the extent that physical site or resource specificity exists as a consequence of a preexisting command economic structure, the existence of competition cognitions is expected to ensure that transaction costs are minimized with hierarchies left intact, thus avoiding "Type II privatization errors." In the following four cases, illustrations of positive and negative examples that have occurred in both low and high specificity situations are provided in the order specified in Figure 4.

Low Specificity Examples

(–) Box 1 – Type I (Collectivization) Error: On-reserve business failures in the Nass River Valley. For the past several years both a gas station and a supply store have operated on-reserve in the Nass River Valley of northwest B.C. The band council has officially and unofficially encouraged these businesses as good for that native community. However, both businesses have failed. According to the individuals familiar with the circumstances, these operations have failed due to insufficient business. Rather than spending their money at these establishments, members of the native community drive by them on the way to larger communities. When we inquired into the reasons for inadequate support, most responses suggested that community members used shopping trips as a form or entertainment and to get out of the relatively isolated community. More specifically, none of the native individuals interviewed felt any loyalty to the success of these business establishments. Here, in a low specificity circumstance (many alternatives to the on-reserve gas or supply store), the result was a collectivization error by both the band council and the business venturers. The error occurred due to inadequate competition cognitions (lack of understanding of market realities, or a lack of the capability to create business strategies that could make the work of the business competitive).

Further, for on-reserve businesses in isolated communities to succeed using collectivization as opposed to competition cognitions, the businesses require the support of the whole community; collectivization must be a total community decision. However, under the present band council system, the council as the central planning authority does not have a consensus-based economic mandate. Rather, it is a majority-vote system (more akin to a market) that consistently produces disaffected minorities who do not necessarily feel bound by council decisions (such as encouraging patronage of private on-reserve businesses). Accordingly, collectivization in market-like (low specificity) settings is predicted to fail. Market alignment, through the effective use of competition cognitions, is predicted to be the key to economic development.

Interestingly, collectivization errors were less likely under the traditional hereditary chief system (as contrasted with the INAC "elected Chief-in-council" system). In our investigation of recorded oral traditions that form the cultural foundation for the peoples of northwest B.C. (Robinson and Wright 1962), the cultural institution that supported effective collective economic activity was "the feast." To ensure that the entire community was in support of collective action in low-specificity circumstances (i.e., where many alternative courses of economic action were present), a feast or a series of feasts was given by the chief, during which consensus for collective action (the requisite competition cognitions) was built. In this manner the hereditary chief system, while still a command economy, was able to ensure that attempts at collective economic action in low specificity circumstances were successful, and thus avoid the Type I (Collectivization) errors encountered in the Nass Valley.

(+) Box 2 – Market Alignment: The heavy equipment operations of Kitamaat Village are privatized. Low specificity circumstances are characterized by the availability of market alternatives (Mitchell 1992; Williamson 1985). Market alignment in low specificity circumstances means that markets are predicted to succeed. Privatization, the movement of non-specific assets from hierarchy to market, is one possibility.

In the case of Kitamaat Village, the band council had jurisdiction over heavy equipment operations, road maintenance equipment, hauling trucks, excavation equipment, etc. Over the past several years, the band council staff, using its continually developing competition cognitions, has begun to systematically contract out these services, gradually transferring the ownership of the equipment, the responsibility, and the related cash flows into the private marketplace.

In this situation specificity is low. Demand for these services is relatively constant. The skills necessary to run these businesses are possessed by a number of individuals who have been trained to effectively and efficiently operate and maintain the needed equipment. Thus, as predicted by Transaction Cognition Theory, for such a market, the transactions related to heavy equipment operations have been successfully externalized from government operations (privatized). The band council selected a course of action that was consistent with the autonomous economizations of the market. The only drawback reported in our field work was the need to go slowly due to the unfamiliarity of decision-makers with this new approach to governance and the inevitable wariness that is created when the outcome is at risk in a new marketplace and is not (as much) under political control.

High Specificity Examples

(+) Box 3 – Hierarchy Alignment: The case of the B.C. First Citizens Fund. In high specificity situations, alignment occurs when governments utilize hierarchy to govern the transaction set in question. In the case of the B.C. First Citizens Fund, this approach has met with success.

In 1969, B.C. Premier W.A.C. Bennett set aside $25 million to be used to fund First Nations' culture, language, and economic development. The fund has grown to over $40 million, according to interviewees familiar with the circumstances. Over the years, the fund has been used as a catalyst for native friendship centres, to provide transportation for native elders to enable the continuation of important cultural institutions, and for student bursaries related to culture, language, and economic development.

According to the predictions of Transaction Cognition Theory, the effective and efficient management of the fund over the years is the result of appropriate competition cognitions. That is, in this high specificity situation (the allocation of funds based on unique projects and judgments about the projects) the existing hierarchy has been essential to the preservation and growth of the fund.

Unfortunately, there are voices in the B.C. political arena who, unaware of the potential for deleterious consequences, suggest that these operations should be privatized. Transaction Cognition Theory suggests privatization of the fund would be unwise, due to the Type II (Privatization) Error that looms if such attempts were to proceed. An example of this error follows.

(–) Box 4 – Type II (Privatization) Error: INAC dismantles existing programs in anticipation of the Nisga'a Treaty signing. On May 11, 2000, the Nisga'a Treaty took effect. Among other things, the treaty provides for lands and money to the Nisga'a First Nation. Although many future decisions of the Nisga'a Tribal Council (now referred to as Nisga'a Lisms Government) may be susceptible to interpretation by Transaction Cognition Theory, it is in the INAC decision framework *prior to the effective date of the treaty* that we see an example of the Type II (Privatization) Error.

During the months and years leading up to the signing of the Treaty, the relationship between the four Nisga'a bands and INAC was less than stable. According to individuals involved in the band governance process, as the treaty process proceeded, the programs on which the bands depended for smooth operations in economic development, health, and education were dismantled well before the need for them expired. If the return of lands and money to the NTC (now referred to as LISMS Government) can be considered privatization, then the premature dismantling and atrophy of existing systems can be described as a Type II (Privatization) Error.

According to our research data, transaction costs increased due to:

- the postponement of capital projects (e.g., subdivision development);

- the inaccessibility of funds for Community Grant Branch Joint Projects (e.g., for recreation centres);
- the refusal to continue Capacity Building (e.g., native leadership development) in an otherwise (to other bands) well-funded program; and
- the curtailment of "off-reserve" services (e.g., community futures).

The comment that summarizes our interviews on these matters during the period leading up to treaty signing is: "if you are Nisga'a, you're shut out of INAC." As a result, economic development, and even broader social development, has been compromised due to the inadequate competition cognitions within INAC that were caused by a privatization error with its roots in the inadequate competition cognitions.

Analysis and Conclusion

In this section, the cases and theory are further analyzed and interpreted to suggest generalizable explanations and concepts for the creation of market economies in northwest B.C. Suggestions for policy and practice are developed. This section is composed of two parts: (1) an analysis of the transaction cognitions currently present and influencing the northwest B.C. situation; and (2) the likely impacts on economic development of the presence or absence of the requisite transaction cognitions.

Analysis of current transaction cognitions

In our meetings with band economic development officers (EDO's) on-reserve, and with other groups of native leaders, we have continually been confronted with the probing question: Why do most private enterprise initiatives on-reserve fail? This article has developed an application of Transaction Cognition Theory that can contribute to answering this question. Transaction Cognition Theory suggests that three sets of effective economic cognitions (planning, promise, and competition cognitions) working together are sufficient for successful economic transacting. In the paragraphs that follow, our analysis of the presence or absence of these three effective cognition sets in the aboriginal case provides some insights.

Under section 20 of the Indian Act of Canada, most transacting must by law[4] involve the band council and the Minister of Indian and

Figure 5: The Planning, Promise, Competition, and Political Cognitions Required in Native Transacting

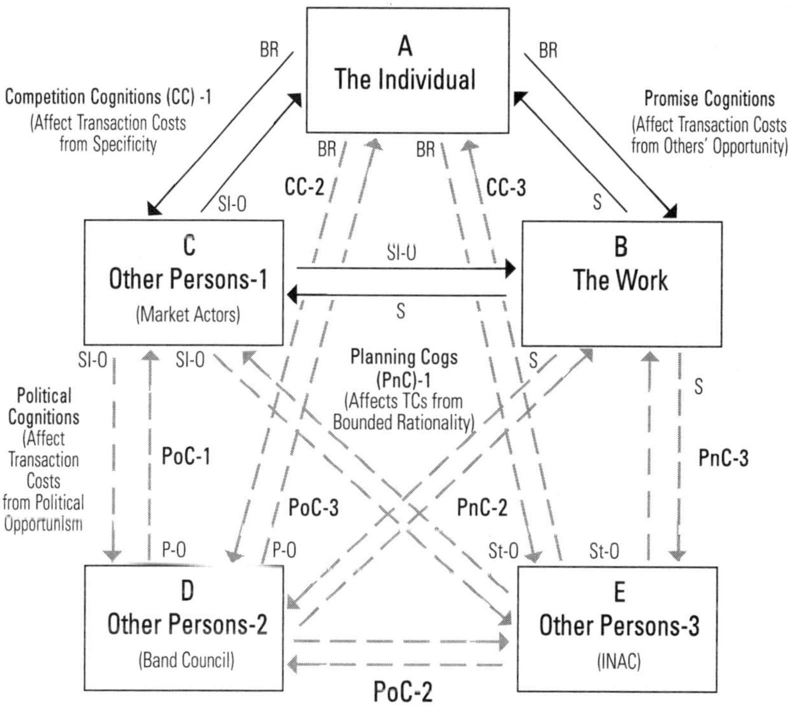

Northern Affairs (the INAC civil servant bureaucracy), in addition to the individual transaction creator, the work, and other persons. Thus, instead of having to master the three cognition sets suggested by Transaction Cognition Theory to be sufficient for market transacting, economic actors in the current on-reserve native economic development arena must master the 10 cognition sets illustrated in Figure 5. To facilitate an explanation of the relationships shown in Figure 5, a discussion of the various components in the diagram follows.

[4] This aspect—the legally mandated inclusion of other parties to a transaction—is critical. Stakeholder theory provides for the systematic identification and explanation of the role of *other persons* who may or may not become involved in a transaction, depending on the relative incentive or disincentive offered within the marketplace. However, when inclusion of certain parties in a transaction is mandated under law, then the model must be altered accordingly to accommodate and explain the cognitions that are required to manage the relationships with mandated parties.

As illustrated in Figure 5, the addition of two "other" parties to the transaction mix greatly complicates the transacting process, and increases the level of cognitive skill needed to succeed within the transacting environment created by the Indian Act. As suggested by previously developed theory, each of the "other" parties to the transacting relationship brings opportunism: types of self-interest seeking with guile. To fully interpret the implications of this higher-complexity transaction relationship, it is useful to first examine the exact nature of the opportunism that each party introduces into the transaction calculus and to explore more deeply the additional transaction cognitions that are implied; and second, to examine in more detail each of the 10 cognition sets required for successful on-reserve transacting.

Implications of multiple additional types of opportunism

The general transaction cognition model suggests that promise cognitions are requisite for dealing effectively with opportunism. Because the on-reserve transacting environment has two additional "other" parties to the transacting process, the problem becomes the identification of the specialized promise models that are needed to support transaction completion. Thus, the introduction of additional theory to support such an explanation is necessary.

The nature of the problem (the evaluation of opportunistic relationships among various groups of others within a transaction) suggests an appeal to political cognition theory, which defines the association of cognitive processes with political (multiple others) behavior (Barner-Barry and Rosenwein 1985, 141). Political cognition theory suggests that political decision-makers also utilize mental models (referred to as operational codes) that are shaped by decision-maker values and political objectives (George 1969; Leites 1951). Thus, political cognition theory suggests a strong relationship between cognition and political behaviours.

However, the political cognition literature appears to be underdeveloped in its explanation of transacting relationships, being "almost deadeningly silent on the issue of leader-follower relationships" (Barner-Barry and Rosenwein 1985, 138). It therefore appears that there is a need for additional theory development in the area of transaction-based political cognitions, especially as these cognitions apply to the dimensionalization of the types and effects of opportunism within the on-reserve transacting system. The analysis proceeds to address this issue next.

The three previously noted "other" parties to the on-reserve transaction are: market actors, the band council, and the Minister/INAC. As expected, market actors (the original other persons in the basic transaction cognition model) bring material (or dollar) opportunism to the transaction. This type of self-interest seeking of goods and services described within the TCE literature (Williamson, 1985, 1991, 1996b) that results in self-protection/hazard minimization behaviors in the face of intrusions (social, political, etc.) into the transacting process. Self-protection cognitions might therefore be considered to be the first political cognition subcategory (PoC-1) of promise cognitions.

With respect to the opportunism introduced by the band council, political cognition theory suggests that the type of opportunism brought into the transacting relationship by elected officials will be based in the acquisition and maintenance of power (Barner-Barry and Rosenwein 1985, 238-239). That is, decision making by the band council is likely to be shaped (at least partially, but materially) by the desire of elected leaders to retain office (power) and exercise authority (Etzioni 1988). Accordingly, the cognitions that are required to manage power opportunism might be characterized as authoritarian cognitions (Barner-Barry and Rosenwein 1985; Stanford 1973, 144), the second political cognition (PoC-2) subcategory of promise cognitions.

The opportunism introduced by INAC involvement is again different. Political cognition theory suggests that the operational codes of a government bureaucracy that is responsible for administering the law (the Indian Act) will center around stewardship – ensuring adherence to the rules inherent in the sociopolitical environment (e.g., the Act) (Rosenberg, Ward, and Chilton 1988, 12) to avoid criticism. Thus, it might be expected that the cognitions that are necessary to manage the stewardship-based interventions of INAC will be compliance-based (PoC-3).

Table 2 presents these relationships in an analytical format similar to that originally utilized to dimensionalize the TCE model (Williamson 1985), and suggests the basic cognition sets necessary for successful transacting. Noteworthy in Table 2 are the systematic relationships among components of the model. Interestingly, the discussion focuses on only one direction of this bidirectional model, as did Williamson's in the case of TCE. For ease of bidirectional reference, Table 2 shows both of the drivers of political relationship promise behaviors (e.g., Altruism ´ Self-protection) implied by the presence or absence of the various types of opportunism that exist within the on-reserve transacting environment. In Figure 6, the political cognition area (containing the specialized native promise models) of the overall model of Figure 5, is illustrated to facilitate theoretical consistency.

The ten types of on-reserve transaction cognitions

As shown in Figure 5, there exist 10 sets of cognitions that, according to the logic of Transaction Cognition Theory, are necessary to accomplish on-reserve economic transactions. Figure 5 utilizes abbreviations to map each type of cognition and the circumstances under which each set is required. In Table 3, each of the 10 types of on-reserve cognition is described in more detail. Given the vastly larger quantity of cognitive maps required to successfully transact in the native economy, each requiring substantial development (Arthur, 1994a; Lord and Maher 1990; Walsh 1995), it is little wonder that few private businesses succeed there. In fact, our analysis suggests that transacting with customers (i.e., the development of private enterprise on-reserve) is the least likely to occur for the reasons that follow.

First, as shown in the case studies and analysis, individual planning cognitions on-reserve are not focused on the customer as the source of economic well-being. Rather, the transaction cognitions of natives on-reserve appear to be focused on the band council as the relevant "other" in the transacting process. The (-) results, such as in the Nisga'a halibut license case, are more prevalent than the uncommon (+) results of the business privatization case of NNBC. Second, as also shown in the case studies, native promise cognitions appear to be stakeholder-based, and quite sophisticated in development; but due to the property rights anomalies created by the Indian Act, the promise cognitions on-reserve are focused away from market transacting and toward band councils and INAC. The (-) situations such as the building on-reserve case are pervasive, while the (+) case of the Sechelt Band property rights is virtually unique. Third, the case studies illustrate the relative absence of market competition cognitions in the aboriginal community in northwest B.C. The cognitions that form the foundation of sophisticated market strategic thinking are, in practical terms, missing: the cases of market alignment are rare, while the cases of collectivization and privatization errors are prevalent. Further analysis of the aboriginal case reveals at least five factors that are likely to contribute to this deficit in on-reserve competition cognitions.

It has long been thought that the absence of a competitive social framework in subsistence-tradition cultures was caused by scarcity, which creates an image of limited good within the minds of the membersof these societies (Foster, 1962). In the native case in northwest B.C.,

Table 2: Some Attributes of Native Political Processes

Source of Potential Conflicts			
(Market Actors) Self-Interest opportunism	(Band Council) Power Opportunism	(INAC) Stewardship Opportunism	Implied Promise: (Type of political relationship behaviour)
0	+	+	Altruism ↔ Self-protection
+	0	+	Laissez-fair ↔ Authoritarian
+	+	0	Service ↔ Compliance

0 = absence, + = presence

Figure 6: Specialized Native Promise Models:
Due to Transaction Costs Resulting from section 20 of the Indian Act

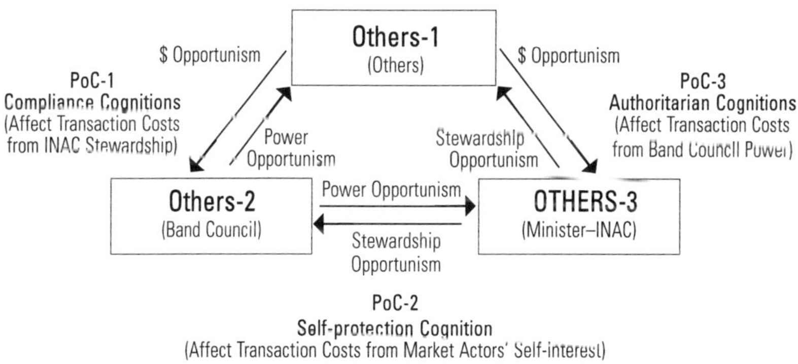

Table 3: Transaction Cognitions Required Due to section 20 of the Indian Act

Transaction Cognitions	Description
Planning Cognitions-1 (PnC)-1	Mental models that assist in developing analytical structure to solve previously unstructured market problems in the provision of the work to those other persons who consume it (e.g., the business plan, which answers the question: What plan is necessary to deliver the work to customers?) (Stevenson, Roberts, and Grousbeck 1994)
(PnC)-2	Mental models that are necessary to ensure band council support of work produced.
(PnC)-3	Mental models that are necessary to ensure that work is approved by/not opposed by, the Minister-INAC.

Transaction Cognitions	Description
Promise Cognitions	Mental models that help in promoting trustworthiness in economic relationships with, e.g., stakeholders (Agle, Mitchell, and Sonnenfeld 1999; Mitchell, Agle, and Wood 1997). Stakeholder identification and salience cognitions (Mitchell and Agle 1997) are essential in market relationships. But see political cognitions (below) for the additional promise cognitions required due to section 20 of the Indian Act.
Competition Cognitions-1 (CC)-1	Mental models that can create sustainable competitive advantage in creator-customer interactions about the work (e.g., I/O strategy: differentiation or cost competitiveness (Porter 1980)).
(CC)-2	Mental models needed to manage creator ↔ band council interactions where there is external power exercised with respect to the work (e.g., Resource Dependence strategy (Pfeffer and Salancik 1978)).
(CC)-3	Mental models needed to manage creator ↔ Minister/INAC interactions about the legitimacy of the work (e.g., Institutional theory-based strategy (DiMaggio and Powell 1983)).
Political Cognitions-1 (PoC)-1	Compliance Cognitions: Mental models needed to manage the relationship between market actors (such as customers) and the band council, in light of the statutory duties of INAC.
(PoC)-2	Self-protection Cognitions: Mental models needed to manage the relationship between the band council and INAC, in light of the self-interest concerns of market actors (such as customers).
(PoC)-3	Authoritarian Cognitions: Mental models needed to manage the relationship between market actors (such as customers) and INAC, in light of the power concerns of the band council.

however, the argument is not as clear-cut. Until the time of first contact with Europeans, native histories chronicle that, apart from the infrequent natural disaster, the society was endowed with plenty (fish, timber, fruits, game, etc.) (Robinson and Wright 1962). Why, then, would competition cognitions be virtually nonexistent in such a society? First, it appears that in the native case in northwest B.C., the image of limited good was less about competition than it was about obedience to the laws of nature. The "law of the land of Ksan" (Robinson and Wright 1962, 5-14), for example, forbids individuals to take more than is needed from nature. Native oral histories are replete with examples of the calamities that befall those who violate this law, such as the morality tale of the Little Goat, and the resulting retribution at "L – La – Matte" (Robinson and Wright 1962, 5-14). Competition cognitions that require the production of surplus for resale may thus be dampened by cultural norms.

Second, the system of government under hereditary chiefs was a command economy that corresponded to its feudal system counterparts found in Western Europe. As suggested in earlier comparisons to the former U.S.S.R., and related work in the small business arena (Mitchell and O'Neil 1998) business conducted under a feudal order is more about politics than about competitive markets. Market competition cognitions are unlikely to be developed or refined under systems that insert politics into transactions.

Third, as indicated in the analysis earlier in this paper (Table 3), the primary source of competition cognitions in modern market transacting is the strategy (Porter 1980). Strategy scholars often note (Thompson and Strickland 1995) that competitive strategy has its roots in military strategy, for example in Sun Tzu's *The Art of War*, from 300 B.C. (trans. Griffith 1984). However, during the years following first contact, the cognitions of warfare were systematically suppressed among native populations to minimize the risk of physical danger to the colonizers. The arts of war – the roots of strategic (competitive strategy) thinking – were thus viewed with great suspicion and repression during the imperialist years (Boldt 1993, 3).

Fourth, within trade-based economies such as those historically present in northwest B.C., most transacting was accomplished using spot markets. As in other more primitive economies, the more sophisticated market mechanisms that involve contracting over space and time, and the supporting competition cognitions, are missing (Olson 1998).

Fifth, the traditional competition cognitions in northwest B.C. that are expected to persist to some degree have been trade and/or tariff-based. Oral histories explain that for vast periods of time the economic framework consisted of the control of territory by clans. The Grizzly Bear clan, for example, exercised its rights of control over territory by control of the Skeena River. Other peoples who desired to utilize the waterway paid tribute to Grizzly Bear chief Neas Hiwas (Robinson and Wright 1962), and by this tariff mechanism, markets such as they existed, were able to operate.

Summary

Thus it appears to us that private enterprise is less likely to occur on-reserve because the focus of native society is not, and has not been, on the market transaction cognition triangle (a b c) depicted in Figure 5. Instead, the focus is on the political cognition triangle (c d c). Thus, on-reserve planning cognitions appear to be misdirected and underdeveloped in the market case. On-reserve promise cognitions also appear to be misdirected and underdeveloped in the market case due to the lack of property rights and the associated presence of an added set of rival promise cognitions, political cognitions. Competition cognitions are essentially missing. This differential is, we believe, the primary reason that on-reserve private business has such difficulty. In the final section of this article we examine the likely impacts on economic development of the present levels and emphasis of transaction cognitions among native peoples in northwest B.C.

Likely impacts on economic development

The presence or absence of the requisite cognitions figures heavily in the actual outcomes that we observe within the native communities of northwest B.C. If William James' assertion that "we become what we think about" is true (1890, paraphrased), and if poverty and economic underdevelopment are about inadequate understanding, as suggested by Grameen Bank founder Muhammad Yunus (1998), and not about land, labour, or capital, then economic development challenges in northwest B.C. can be productively viewed through the lens of Transaction Cognition Theory.

Our research shows that the presence or absence of the requisite cognitions is not uniform across the native communities of northwest B.C. Rather, as illustrated, there are pockets of cognitive capability that exist along side an absence of the transaction cognitions: planning, promise, and competition. Thus, the systematic interpretation of the results of our research suggests that the three-variable analysis model

Figure 7: Example Society Level Variations in Economic Outcome Condition as a Function of Planning, Promise, and Competition Cognitions

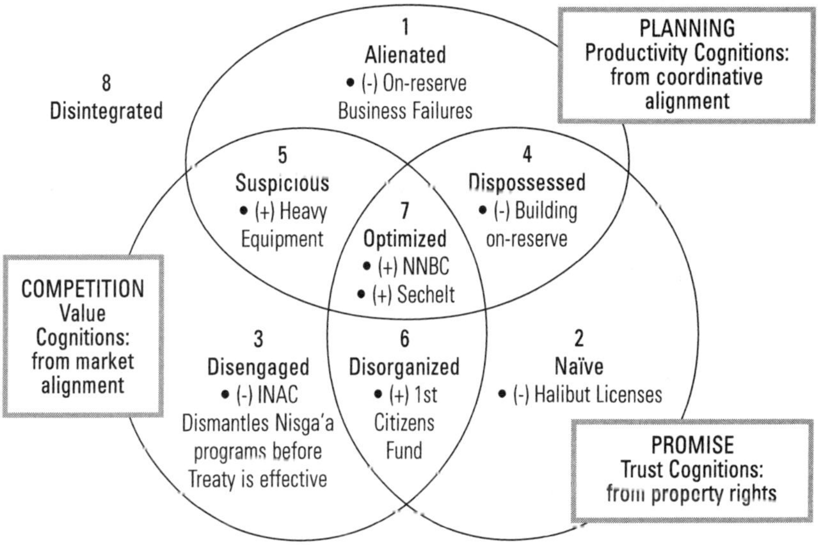

applies (Figure 7). When examining the process of economic development in these communities, it appears prudent to become aware of the various circumstances that form the cognitive foundation on which the next steps must be built. As illustrated in Figure 7, a variety of economic and cognitive consequences of native planning, promise, and competition cognitions that presently exist in northwest B.C. can be expected.

The adjectives utilized as identifiers within the diagram describe eight possible outcomes of the presence or absence of the three Transaction Cognition Theory variables. Within each area, we have plotted the example cases utilized in this paper as illustrations. The relationships and illustrations shown in Figure 7 suggest that the positive-negative economic consequences of the presence or absence of particular society-level cognitions are systematic. As a result, the likely impacts on economic development of the possession of the three key transaction cognitions become clearer.

Conclusion

Our conclusions are relatively straightforward. It is our assessment, based on the concepts and analysis presented in this paper, that two contingencies will influence the likely success of economic development initiatives in northwest B.C. First, unless on-reserve property rights akin to those available to other transacting parties in the global economy are made available, economic development initiatives are predicted to fail, as they have an inadequate foundation in a market economy, especially with the corresponding retreat of band councils and INAC from their complicating role in transacting. It is unclear to us whether ongoing treaty negotiations with various First Nations will produce such rights or whether additional governance options should be considered. It is clear that, from the standpoint of success in economic development, this objective should have the highest priority with negotiators on both sides of the table.

Second, unless economic decision-makers (leaders, venturers, and even the general members of on- and off-reserve native society) possess levels of planning, promise, and competition cognitions that are sufficient for their roles in economic development, the settlement of aboriginal claims based on land and cash will be inadequate to ensure a vibrant on-reserve economy. Given the brief elements of history that we have incorporated in our research, and the evidence reported in our case studies as a sketch of the transacting landscape in northwest B.C., it is our assessment that a significant and long-term commitment to the acquisition and maintenance of modern market cognitions is essential, if the economic aspirations of aboriginal peoples, that we have heard expressed are to be realized.

Thus, our analysis boils down to two simple dicta: property rights and transaction cognitions are the basis for sustainable economic development within the aboriginal communities of northwest B.C. If, as William James suggests, we become (or realize) what we think about, then these are the two objectives that, in our opinion, are worth thinking about.

Acknowledgments

The authors gratefully acknowledge the assistance of the Skeena Native Development Society First Nations Think Tank on Wealth Building, the research endowment of the Francis G. Winspear Chair in Public Policy and Business and the International Centre for Venture Expertise ant the University of Victoria, Fritz Faulhaber, and Charmaine Stack for their research support. Further, we thank the editors, reviewers, and other conference participants for their helpful comments on this paper. We are also grateful to our families and colleagues for their inspiration and support.

References

Agle, B.R., Mitchell, R.K., and Sonnenfeld, J.A. (1999). Who matters to CEOs? An investigation of stakeholder attributes and salience, corporate performance, and CEO values. *Academy of Management Journal Special Research Forum on Stakeholder Theory* 42: 507-25.

Arrow, K. J., ed. (1969). *The Organization of Economic Activity: Issues Pertinent to the Choice of Market Versus Nonmarket Allocation.* Washington.: U.S. Government Printing Office.

Arthur, W. B. (1994a). Complexity in economic theory: Inductive reasoning and bounded rationality. *AEA Papers and Proceedings* 84(2): 406-11.

Arthur, W. B. (1994b). *Increasing Returns and Path Dependence in the Economy*. Ann Arbor: University of Michigan Press.

Barner-Barry, C. and Rosenwein, R. (1985). *Psychological Perspectives on Politics*. Englewood Cliffs, New Jersey: Prentice-Hall.

Barney, J. B. and Hansen, M. H. (1994). Trustworthiness as a source of competitive advantage. *Strategic Management Journal* 15: 175-90.

Boldt, M. (1993). *Surviving as "Indians"*. Toronto: University of Toronto Press.

Csikszentmihalyi, M. (1988). Society, culture, and person: A systems view of creativity. In *The Nature of Creativity*, ed. R.J. Sternberg, 325-39. New York: Cambridge Press.

DiMaggio, P. J. and Powell, W. W. (1983). The iron cage revisited: Institutional isomorphism and collective rationality in organizational fields. *American Sociological Review* 48: 147-160.

Etzioni, A. (1988). *The Moral Dimension*. New York: Free Press.

Foster, G. M. (1962). *Traditional Cultures and the Impact of Technological Change*. New York: Harper Row.

Gardner, H. (1993). *Creating Minds*. New York: Basic Books.

George, A. L. (1969). The "operational code": A neglected approach to the study of political leaders and decision-making. *International Studies Quarterly* 13: 190-222.

James, W. (1890). *The Principles of Psychology*. New York: Holt.

Kirzner, I. M. (1982). The theory of entrepreneurship in economic growth. In *Encyclopedia of Entrepreneurship*, ed. C. A. Kent, D. L.

Sexton and K. H. Vesper, 272-76. Englewood Cliffs, New Jersey: Prentice-Hall.

Kornai, J. (1986). The Hungarian reform process: Visions, hopes, and reality. *Journal of Economic Literature* 24: 1687-1737.

Lange, O. (1938). On the theory of economic socialism. In *On the Economic Theory of Socialism*, ed. B. Lippincott, 55-143. Minneapolis: University of Minnesota Press.

Leites, N. (1951). *The Operational Code of the Politburo*. New York: McGraw-Hill.

Lord, R. G. and Maher, K. J. (1990). Alternative information-processing models and their implications for theory, research, and practice. *Academy of Management Review* 15: 9-28.

Mitchell, R. K. (1992). From hierarchy to market: Mitigating societal impacts of governance transition in the macro institutional setting. In *International Association for Business and Society Proceedings*, ed. S.A. Waddock, 232-40. Leuven, Belgium.

Mitchell, R. K. (1999). A transaction cognition theory of entrepreneurship. Working Paper, University of Victoria.

Mitchell, R. K. (2001). *Transaction Cognition Theory and High Performance Economic Results*. Victoria: International Centre for Venture Expertise.

Mitchell, R. K. and Agle, B. R. (1997). Stakeholder identification and salience: Dialogue and operationalization. In *International Association for Business and Society 1997 Proceedings*, ed. J. Weber and K. Rehbine, 365-70. Destin, Florida.

Mitchell, R. K., Agle, B. R., and Wood, D. J. (1997). Toward a theory of stakeholder identification and salience: Defining the principle of who and what really counts. *Academy of Management Review* 22: 853-86.

Mitchell, R. K. and O'Neil, S. J. (1998). Feudal institutions and small business governance: An exploration and comparison. *Journal of Small Business and Enterprise Development* 5(3): 208-18.

Mitchell, R.K., Smith, B., Seawright, K.W., and Morse, E.A. (2000). Cross-cultural cognitions and the venture creation decision. *Academy of Management Journal*, 43: 974-93.

Neisser, U. (1967). *Cognitive Psychology*. New York: Appleton-Century-Crafts.

Olson, M. (1998). *Power and Prosperity: Outgrowing Communist and Capitalist Dictatorships*. New York: Basic Books.

Pearce, J. L. (1991). From socialism to capitalism: The effects of Hungarian human resources practices. *The Executive* 5(4): 75-88.

Pfeffer, J. and Salancik, G. (1978). *The external control of organizations: A resource dependence perspective*. New York: Harper and Row.

Poelzer, G. (1998). Land and resource tenure: First Nations traditional territories and self-governance. In *Prospering Together: The Economic Impact of the Aboriginal Title Settlements in B.C.*, ed. R. Kunin, 85-110. Vancouver: The Laurier Institution.

Porter, M. E. (1980). *Competitive Strategy: Techniques for Analysing Industries and Competitors*. New York: Free Press.

Porter, M. E. (1985). *Competitive Advantage*. New York: Free Press.

Read, S. J. (1987). Constructing causal scenarios: A knowledge structure approach to causal reasoning. *Journal of Personality and Social Psychology* 52: 288-302.

Ritzer, G. (1979). Toward an integrated sociological paradigm. In G. Simzek (Ed.), *Contemporary issues in theory and research*: 25-46: Westport, CT: Greenwood.

Robinson, W. and Wright, W. (1962). *Men of Medeek*. Terrace, B.C.: Northern Sentinel Press. Original translation done in 1937, but not published until 1962.

Rosenberg, S. W., Ward, D., and Chilton, S. (1988). *Political Reasoning and Cognition*. Durham: Duke University Press.

Rousseau, D.M. (1985). Issues of level in organizational research: Multi-level and cross-level perspectives. *Research in Organizational Behavior*: 1-37. Greenwich, CT: JAI Press.

Schein, E. H. (1985). How culture forms, develops, and changes. In *Gaining Control of the Corporate Culture*, ed. R. H. Kilmann, M. J. Saxton and R. Serpa, 17-43. San Francisco: Jossey-Bass.

Simon, H. A. (1979). Rational decision-making in business organizations. *The American Economic Review* 69: 493-513.

Stanford, N. (1973). Authoritarian personality in contemporary perspective. In *Handbook of Political Psychology*, ed. J.N. Knutson. San Francisco: Jossey-Bass, Inc.

Stevenson, H. H., Roberts, M. J., and Grousbeck, H. I. (1994). *New Business Ventures and the Entrepreneur*. Homewood, Illinois: Irwin.

Teece, D. J. (1986). Profiting from technological innovation. *Research Policy* 15: 285-305.

The Indian Act (1876). Department of Justice Canada Website,

Thompson, A.A., Jr. and Strickland, A.J, III (1995). *Strategic management*. Chicago: Irwin.

Tzu, S., and Griffith, S. (1984). *The Art of War*. London: Oxford University Press.

Walsh, J. P. (1995). Managerial and organizational cognition: Notes from a trip down memory lane. *Organization Science* 6: 280-321.

Williamson, O. E. (1975). *Markets and Hierarchies*. New York: Free Press.

Williamson, O. E. (1985). *The Economic Institutions of Capitalism*. New York: Free Press.

Williamson, O. E. (1991). Strategizing, economizing, and economic organization. *Strategic Management Journal* 12(Summer): 75-94.

Williamson, O. E. (1996a). Economics and organization: A primer. *California Management Review* 38(2): 131-143.

Williamson, O. E. (1996b). *The Mechanisms of Governance*. Oxford: Oxford University Press.

Yunus, M. (1998). Poverty alleviation: Is economics any help? Lessons from the Grameen bank experience. *Journal of International Affairs* 52(1): 47-57.

Chapter 7

The Link Between Innovation and the Use of Human Resources in British Columbian Enterprises

J.A.D. Holbrook and L.P. Hughes

Abstract

Virtually all studies of innovation have focused on economic impacts. However, innovations have impacts on society and the development of human capital that are at least as important as the economic impacts. Policymakers need to consider the application of science and technology to, and the role of technological innovation in, the development of human capital at the firm level. Because regional clusters are the building blocks of the Canadian system of innovation, this study focuses on the relationship between human resources and innovation in a region of British Columbia. The results suggest that human resource practices and the human capital needs of innovative firms differ from non-innovative firms. The findings also suggest that the manner in which innovation is defined – "new to the firm" versus "new to the market" – influences the findings of innovation studies. Implications for policy and research are discussed.

Introduction

There is often a perception that the primary objective of science and technology (S&T) policy is the further development of industries in the manufacturing sector. Governments in most jurisdictions support, in one way or another, S&T programs in the firm belief that investments in S&T have a positive, if indefinable, effect on economic growth. It is widely assumed that economic growth is a social benefit, and that growth in knowledge and technological inputs must inevitably result in social progress. While virtually all studies of innovation have focused on economic impacts, innovations have impacts on society and on the development of human capital which are at least as important as their economic impacts. Policymakers need to consider the application of S&T to, and the role of technological innovation in, the development of human capital at the firm level.

Studies of innovation in Canada have been carried out at the national level, but because of the preponderance of industrial activity in Ontario and Quebec, the results understandably reflect the characteristics of these manufacturing-based provinces (e.g. Baldwin and Da Pont 1996; Baldwin, Chandler, Le, and Papailiadis 1994). There have also been studies of regional industrial clusters (or "poles") and comparisons of regional, or sub-national, innovative performance. A recent review of this subject, in the Canadian context, has been published by de la Mothe and Paquet in *Local and Regional Systems of Innovation* (1998). These regional clusters are the building blocks of the Canadian national system of innovation. But in the knowledge-based economy, where knowledge, embedded in the training of the human capital of the innovative firm, is the primary resource for the innovative firms, what is the relationship between the innovative behavior of the firm and the way it manages its human resources? Is it possible that there might be regional variations?

Survey Methodology

A short questionnaire for use with B.C. enterprises was developed by the authors at the Centre for Policy Research on Science and Technology (CPROST) at Simon Fraser University and the overall results reported (Holbrook and Hughes 1998). The questionnaire was not intended to cover all aspects of technological innovation identified in the OECD *Oslo Manual* (1997), but it had to conform to the main points in the OECD standard. To ensure a reasonable response rate, the questionnaire was short (no more than one page, printed on both sides) so that it would be user-friendly, take little managerial time to complete, be comprehensible to small technology-based entrepreneurs based in B.C., and be faxable to expedite its return.

A similar survey covering the Okanagan region of B.C., a non-metropolitan area dependent mainly on agriculture and resource extraction, was conducted in July 1997 (Holbrook, Hughes, and Finch 1999). For the Okanagan phase of the project, the questionnaire was modified from the version used for the Lower Mainland. Some of the modifications were the result of conclusions drawn from the analysis of the Lower Mainland responses; others were added to provide additional information on knowledge management and highly qualified personnel.

The samples in both surveys were drawn from two industrial sectoral groups, "high technology" and "policy sectors". In eight industrial sectors, firms were selected from the two groups: manufactured products, computer services, and technical services (high tech), and food, forest, and electrical products, construction, and transportation (policy). To be included in the sample, a firm had to employ at least five people and have been in operation for at least five years. These criteria were applied based on the assumption that small, new firms are still in the process of stabilizing, and are likely to be quite transient. It is commonly thought that much innovation occurs in this entrepreneurial environment, the so-called "bleeding edge." These surveys were focused on the characteristics of *successful* innovators. Given that the majority of new ventures fail within the first five years, firms that survived the first five years are more likely to be the successful innovators.

From this survey of innovation of enterprises in four sectors, it was possible to select questions that related innovation to human capital management practices. To no one's great surprise, innovative firms (those that had introduced a new product or process within the previous five years), were also those that scored highly on questions that measured their ability to manage highly skilled human resources.

Innovation in a Regional Market

In the innovation literature, firms are described as being innovative if they have developed a technologically new product or process. The *Oslo Manual* defines an innovative firm as one that has introduced a product or process innovation that is "*new to the firm*" in the past *three* years. The *Green Paper on Innovation* issued by the European Commission (1995), which used results from *Oslo Manual* type surveys in Europe suggested that over half of European firms are innovative and that the larger the firms, the more innovative they are. This result appears to be the same as the CPROST results. However, during focus group testing of the CPROST surveys in B.C., the participants were emphatic that there were significant problems with both the three-year time frame for innovation and its acceptance of "*new to the firm*" as being innovative. Thus, in the process of carrying out the surveys, CPROST changed two component variables of the construct "innovation".

The first change extended the period for product introduction from three years to five years, thereby lowering one of the entry requirements for innovation. Except in some highly competitive industrial sectors (such as computers and software) product life cycles are not usually three years or less. In most manufacturing industries, as well as in most services, product life cycles are at least three years, and sometimes much longer. Given a three-year innovation criterion, many producers of large capital goods, ranging from aircraft to automobiles, would only be considered innovative for about three years of any given decade, given their approximately ten-year product development cycles.[1] The problem is even worse in the resources sector. Most resource-based industries innovate rarely, and when they do, it is through the introduction of new processing equipment rather than through innovations in their product lines. Furthermore, size does matter: the problem is ameliorated in the case of large firms by having multiple product lines and staggering product life cycles, while small and medium-sized firms with single product lines are unable to demonstrate such consistent innovation. This result is shown in the Canadian 1993 Survey of Innovation in Manufacturing Enterprises (Baldwin and Da Pont 1996). For the period 1989-1992, 50% of larger firms were considered innovative, compared to only 30% of small firms (fewer than 100 employees).

In a region such as British Columbia, with a heavy dependence on the resources sector and a large number of small and medium-sized enterprises, this problem of product life cycle had to be addressed. Using the five-year development cycle, the CPROST surveys found 75% to 80% of firms to be *potentially* innovative.

The second CPROST change addressed the issue of defining the market in which the innovations occur, and had the effect of moving the entry requirement in the opposite direction. To most people in the working world developing and selling new products, and competing with one another, innovation has quite a specific meaning. "New" is not equivalent to "innovative." Anyone can introduce a product they have not sold before or a production process they have not used before, particularly after a competitor has taken the risks and proven the new technology. Innovation to these people implies a large element of risk-taking, of putting the company on the line to become a market leader, rather than following simply for the sake of survival. To be innovative, a product should have no equivalent competition at the time it is introduced, therefore being *new to the market*. This is a customer-based definition of

[1] The *Oslo Manual* defines a technologically new product as one whose characteristics "differ significantly from those of previously produced products." While this may be a subjective definition, it clearly rules out many of the incremental changes that manufacturers of large capital items incorporate in their product lines.

innovation, and it is the purchasing habits of customers that determine whether an innovation is successful or not.

Using Schumpeterian arguments, market instability, and consequently growth, comes about with changes to the productive process, and the producer drives this process of innovation. Nevertheless, it is incumbent on the consumer to accept the new product and, by purchasing it, to encourage and reward the innovator. It is the first new product in the market that introduces the instability that causes growth, while subsequent entries by competitors attempt to restore the stability of the market and eliminate the possibility of growth until the next innovation comes along. Consequently, "*new to the firm*" should not be considered the entry point for innovation; indeed, most of the time, it is exactly the opposite, restoring the stability to an economy destabilized by innovators. But these processes happen at a local or regional level; an innovator may well be the first entrepreneur to introduce a product or process to a region, particularly if the regional economy is not manufacturing-based and the technology is imported. Clearly the quality of people in the firm is key to innovation at the regional level. Their regional outlook – culture, if you will – is a big factor in their innovative behavior. In regional economies, skilled personnel are at a premium; regional firms must compete with larger firms in bigger economies for highly trained individuals, and often cannot meet the market levels for their salaries. Thus CPROST coupled the longer life cycle with the "*new to the market which you serve*" concept to arrive at a definition of innovation for a regional market, such as B.C. (abbreviated as "*new and unique*"). In doing so, we were attempting to seek out the entrepreneurs who destabilized the regional, or in the case of the Okanagan Valley, local market, and to find out how they differed in their management of highly skilled personnel.

Survey Results

The results below summarize the responses to questions on the innovation surveys that refer to personnel practices within the firms, and which if answered positively would be indicative of good personnel management within the firm.

Training programs

Innovative firms are more likely to have some type of training programs than non-innovative firms (Table 1).

Table 1: Training Programs by Innovativeness

		Training Programs		
		No	Yes	sig of χ^2
New Product or Process in the past five years (% of respondents by category)	No	5%	11%	0.074
	Yes	15%	70%	
New and Unique Product or Process in the past five years (% of respondents by category)	No	13%	35%	0.028
	Yes	8%	45%	

Training was carried out by firms of all sizes, but larger firms were more likely to have training programs: 77% of small firms, 85% of medium firms, and 91% of large firms reported having training programs. However, there appears to be a strong linkage between training and the "*new and unique*" factor, and the presence or absence of training programs, a linkage that is less strong when the standard Oslo definition of "innovative" is used.

Acquisition of skills

Acquisition of skills is always an important issue for firms. Firms were asked how they went about acquiring skilled workers (Tables 2 and 3).

Availability of personnel

The availability of personnel is always a concern for managers and owners. Respondents were asked whether the availability of qualified personnel helped, had no effect, or hindered innovation in their firms (Tables 4 and 5).

Using the Oslo definition of innovation, there is no significant relationship present in the data. However, by using the "*new and unique*" definition, a significant relationship is indicated between the effects of availability of personnel and innovativeness. The data shows that for innovative firms, the availability of qualified personnel helps innovation, where it is considered to have an effect.

Post-secondary education

Specific skills and a demonstration of the ability to learn are two major characteristics of individuals with post-secondary education, be it from

Table 2: Skill Acquisition Strategies by Innovativeness

		Train to acquire req'd skills	Hire to acquire req'd skills	Contract to acquire req'd skills
New Product or Process in the past five years (% of respondents by category)	No	80%	77%	50%
	Yes	91%	75%	45%
New and Unique Product or Process in the past five years (% of respondents by category)	No	87%	69%	31%
	Yes	92%	80%	54%

a technical college or a university. Approximately 40% of the working population of British Columbia has some post-secondary education.[2] Thus, firms were divided into two categories: those with more than 40% of employees with some post-secondary education and those with less than 40% (Tables 6 and 7).

There is a significant relationship between the number of employees with post-secondary education and whether the firm has produced an innovation that is "*new and unique.*" The relationship between these two variables using the Oslo definition of innovation is not significant. Interestingly, there is no link between having a high percentage of employees with post-secondary education and training programs in the firm. The presence of well-educated employees does not appear to automatically create a demand for further training. These findings are a strong argument for additional public sector investment in post-secondary education programs, and for enhancing access to these programs.

Other Results

Statistics Canada has recently carried out a major survey of innovation in the manufacturing and natural resources sector.[3] As part of this survey, it asked questions concerning training and hiring practices in Canadian industry at the national level. This survey used the *Oslo Manual* (OECD 1997) definition of "*new to the firm*" innovation. They found the following:

[2] British Columbia has a higher percentage of workers with post-secondary training (37%) than the national average (33%).

[3] The questionnaire is available on the Statistics Canada Web site at http://www.statcan.ca/english/concepts/pdf/science/0497-99.pdf.

Table 3: Skill Acquisition Strategies by Firm Size

	Train to acquire required skills	Hire to acquire required skills	Contract to acquire required skills
Small (less than 20 employees) (% of respondents by category)	89%	70%	44%
Medium (less than 100 employees) (% of respondents by category)	93%	86%	46%
Large (more than 100 employees) (% of respondents by category)	81%	88%	50%

Obstacles to innovation

- 62% found it difficult to hire qualified staff and workers.
- 33% found it difficult to retain qualified staff and workers.
- 62% found it difficult to devote staff to innovation.
- 40% reported a lack of skilled personnel.
- 18% reported a lack of marketing capability.

Factors assisting innovation

- 24% assigned a high level of importance to hiring university graduates.
- 40% assigned a high level of importance to hiring graduates from technical schools and colleges.
- 68% assigned a high level of importance to hiring experienced employees.
- 10% assigned a high level of importance to recruiting from abroad.
- 80% assigned a high level of importance to training employees.
- 60% assigned a high level of importance to using employee teams.

Table 4: Effect of Availability of Personnel by Innovativeness

		Availability of Personnel			
		Hinders	No Effect	Help	sig of χ^2
New Product or Process in the past five years	No	32%	28%	40%	0.396
	Yes	21%	39%	40%	
New and Unique Product or Process in the past five years	No	27%	42%	30%	0.043
	Yes	19%	33%	48%	

Analysis and Conclusions

A "*new to the market which you serve*" measure of innovation presents a methodological concern for many innovation researchers, which lies in the problem of defining and operationalizing the term "market." (see Holbrook and Hughes 2001). This study made no attempt to precisely determine the market of each respondent; it assumed that the respondent knew the market in which his or her firm competed. Strictly speaking, this lack of definition introduced a degree of uncertainty: how do we know that we mean the same thing by the term "market" as did the respondent? However, this perception of uncertainty may be to some degree a case of academic isolation from the practitioners being studied. To a business person, "market" has a specific meaning: the set of all potential customers for a firm's products and/or services. This use of the word is more akin to its use in "market share," referring to that portion of all potential customers using a particular product, than to the more general meaning of the term "market" (as in "free market" or "market failure") to economists.

Using a "*new to the market which you serve*" measure addresses other problems inherent in innovation surveys based on the *Oslo Manual*. According to the Manual, innovations fall into one of three categories: "*new to the firm,*" "*new to the nation,*" "*new to the world.*" The "*new to the firm*" category has already been addressed. A product that is "*new to the world*" is obviously innovative, although it prompts the question "How do you know?" or the statement "Prove it!" Only a very small percentage of new products are "*new to the world.*" These innovations are of great interest, since they indicate extreme competence of the firms and of the systems of innovation producing them. That leaves "*new to the nation.*" Although this measure is perceived to be of importance to policy

Table 5: Effect of Availability of Personnel by Firm Size

	Availability of Personnel		
	Hinders	No Effect	Help
Small (less than 20 employees) (% of respondents by category	21%	38%	41%
Medium (less than 100 employees) (% of respondents by category)	34%	32%	34%
Large (more than 100 employees) (% of respondents by category)	9%	45%	46%

makers, it does not necessarily represent the reality of practitioners, who are more concerned with their markets. Their markets may be regional, or they may be transnational; markets rarely coincide with national boundaries, except in cases of highly regulated or protected industries.

As this research program has found, firms serving non-metropolitan regional markets tend to have low exports and rely on suppliers and customers as sources of innovation. These firms import knowledge to a region. On the other hand, firms serving transnational markets export products or services beyond their regional or national milieu and rely on internal R&D as a source of innovation. These firms are knowledge exporters. The importing or exporting of knowledge can serve as an excellent indicator of the expertise of a regional system of innovation. This indicator is also very useful when applied to particular industrial sectors or clusters, since it can signal the growth of pockets of expertise, or clusters. A predominance of firms with a regional focus, importing knowledge, will indicate the region is underdeveloped. On the other hand, a region or cluster dominated by firms with a transnational focus will indicate competence or even special expertise in that region.

Some messages for policy makers interested in the emerging knowledge-based economy can be derived from the data. Innovative firms do appear to need trained individuals, and those that have produced innovations which are "*new to the market which you serve*" need them even more so. Governments, particularly those in resource-based economies striving to make the transition to the knowledge-based economy, should consider increasing their investments in post-secondary education. According to the OECD (2001), several of the economies in the European Union have increased their per student expenditures in post-secondary education, while in Canada this expenditure has actually fallen since 1990.With the current emphasis on job creation as a policy goal

Table 6: Employees with Post-secondary Education by Innovativeness

		Employees with Post-secondary Education		
		< 40%	> 40%	sig of χ^2
New Product or Process in the past five years (% of respondents by category	No	61%	39%	0.250
	Yes	48%	52%	
New and Unique Product or Process in the past five years (% of respondents by category)	No	61%	39%	0.008
	Yes	40%	60%	

in itself, the analysis of non-high tech sectors becomes more important. Natural resource-based industries and consumer service-based industries can all be innovative within their markets. In B.C., these service industries tend to cluster by sector, so that it is important to be able to situate them in any policy framework devoted to enhancing the innovativeness of firms as a whole. The link between the tourism sector and other (innovative) sectors such as agrifoods is also important, at least in the B.C. context.

While the limited data from the survey can only provide a glimpse of the policy issues emerging from the analysis of regional results within B.C., the effects of geographical separation do appear to influence the responses. In previous surveys of high technology firms in the Okanagan, it was reported that lifestyle is an important component for firms choosing to locate there and for employees being drawn to the region (Padmore, private communication 1997). It may be that in some sectors, innovative firms may succeed in remote areas, simply because of the temperament of the individuals who are likely to work in those sectors. The question remaining is whether or not there is a clustering effect. Is there a lower limit to the number of highly skilled individuals required to establish an innovative cluster, or is a cluster of individuals, or firms, required to establish an innovative industrial sector? Do innovative firms attract skilled individuals, or the reverse, or is it a matter of establishing an environment in which both skilled individuals and innovative firms can flourish? Perhaps it is a case of "If you build it they will come"?

Another area requiring improvement is knowledge on the level of skill transfer from studies to employment. Given the high cost of post-secondary education, more knowledge is needed as to how the resulting talents are used, and how, over time, technical knowledge is either

Table 7: Employees with Post-secondary Education by Firm Size

	Employees with Post-secondary Education		
	< 40%	> 40%	sig of χ^2
Small (less than 20 employees) (% of respondents by category	45%	55%	0.037
Medium (less than 100 employees) (% of respondents by category)	61%	39%	
Large (more than 100 employees) (% of respondents by category)	80%	20%	

augmented or depreciated. Studies of the stocks and flows of human capital lead directly to the study of the actors and networks that make up a national system of innovation.[4] This is a field that is only just beginning to be examined, but which is probably more important in smaller economies than in larger ones, where the sheer number of networks and individual actors results in individual actor-networks having less individual influence on the system.

Finally, given the strong correlation between the positive responses to the personnel-related questions, and the "*new and unique*" definition of innovation, it appears that policy analysis should be directed towards this definition of innovation . As noted above, respondents are probably better equipped than researchers to define what their markets are and whether their innovations are indeed new to the markets which they serve. Firm innovativeness should be based on a "*new and unique*" determination. "*New to the firm*" is not necessarily innovative, and "*new to the nation*" does not address the economic realities of regional or transnational markets. "*New to the world*", while capturing innovation, does not capture all innovative activity within a regional or a national system of innovation. Innovation research, to capture innovative activity in most manufacturing and service sectors, must also capture data on product life cycles and analyze that data by industrial sector. This analysis would allow innovative activity to be determined on the basis of actual industrial conditions, rather than an arbitrary, and possibly inappropriate, external determination of product life cycle.

[4] See, e.g., the discussion of national systems of innovation published by the OECD (1997).

Acknowledgments

The authors would like to acknowledge the encouragement and support for this project from the Social Sciences and Humanities Research Council of Canada, Statistics Canada and the Science Council of British Columbia. Some of this material has appeared in a thesis prepared by Hughes (1999) as part of the requirements for an M.A. degree at the School of Communication at Simon Fraser University.

References

Baldwin, J., and Da Pont, M. (1996). Innovation in Canadian Manufacturing Enterprises. cat. no. 88-513-XPB. Ottawa: Statistics Canada.

Baldwin, J. R., Chandler, W., Le, C., and Papailiadis, T. (1994). Strategies for Success. cat. no. 61-523-RPE. Ottawa: Statistics Canada.

de la Mothe, J. and Paquet, G., ed. (1998). *Local and Regional Systems of Innovation.* Boston: Kluwer Academic Press.

European Commission (1995). *Green Paper on Innovation.* Brussels/Luxembourg: European Commission

Holbrook, J. A. D. and Hughes, L.P. (1998). Innovation in enterprises in British Columbia. In *Local and Regional Systems of Innovation,* ed. J. de La Mothe and G. Paquet. Boston: Kluwer Academic Press.

Holbrook, J. A. D., Hughes, L.P., and Finch, J.A. (in press). Characteristics of innovation in a non-metropolitan area: The Okanagan Valley of British Columbia. In *Systems and Policies for the Globalized Learning Economy,* ed. P. Conceicao, D. V. Gibson, M. V. Heitor, and C. Stolp. Westport, Conn.: Greenwood.

Holbrook, J.A.D. and Hughes, L.P. (2001) Comments on the use of the OECD Oslo Manual in non-manufacturing based economies@, *Science and Public Policy* 28(2).

Hughes, L.P. (1999). *Regional Systems of Innovation: Theory and practice.* Masters thesis, School of Communication, Simon Fraser University, Vancouver, British Columbia.

Organization for Economic Cooperation and Development (OECD) (1997). *Proposed Guidelines for Collecting and Interpreting Technological Innovation Data (The Oslo Manual).* 2nd ed. Paris: OECD.

Organization for Economic Cooperation and Development (OECD) (1997) *Managing National Systems of Innovation.* Paris: Committee for Science and Technology Policy, OECD.

Organization for Economic Cooperation and Development (OECD) (2001) *The OECD in Figures.* Paris: The OECD Observer.

Padmore,T. (1997). Private communication, quoting studies carried out by Impax International for the Regional District of the Central Okanagan, Kelowna, British Columbia.

Chapter 8

Networking Support to Innovative Young Firms: Pathways of Knowledge

Cooper H. Langford, Anne Tyrie, and Janet Peace

Abstract

The economic advantages of networking and strategic partnership are well known. This paper attempts to go beyond the practical comments to interpret the process of knowledge flow that takes place in networks. It is especially important to identify what networking processes can be made available to small, innovative, high technology firms. How do we unleash the broadest scope of knowledge in support of innovation? In response to this question, this study focuses on the institutional components of the regional innovation system in Alberta.

Introduction

Networking is important in the evolution of young, high growth, technology-based firms. From a review of earlier literature and case studies, we developed six propositions about networking that are phrased in a language of practice (see Table 1). These served as the basis for conversations with executives from three firms that belong to the class under study, and two executives of organizations that promote networking. Lessons from the MettNet Institute, with which we have all been involved, and the experience of one of us with Calgary agencies, are the final elements of our sources. The interviews served to test the degree to which the propositions captured the experience of active players. The propositions tended to focus discussion on aspects of networking that reflect participation in regional innovation systems.

The economic advantages of networking and strategic partnership are well known. This paper attempts to go beyond the practical comments to interpret the processes of knowledge flow that take place in

Table 1

1. **SME's use networking as a source of new knowledge and will increasingly do so.**
   Firms today cannot contain all necessary knowledge within themselves. Growth SME's currently make extensive use of outside resource services and plan to increase their linkages to such sources of expertise significantly in the future. (Networks of Innovators 1990)

2. **Networking sustains innovation.**
   Networking bears a strong association with growing and innovative firms. The latter develop networks as a means of coping with the liabilities of newness. (Networks of Innovators 1990)

3. **Networks are multidimensional relationships.**
   Networking conducive to innovation is best approached not as a structure, but as the development of multidimensional relationships that go beyond explicit knowledge sharing to long-term partnership activity, involving the exchange of implicit (tacit) knowledge. (Steier and Greenwood 1999)

4. **Several types of organizations sustain SME networks.**
   Our investigations suggest that the types of organizations supporting SME growth and innovation include both networking and direct service organizations. Networking organizations provide the linkages between SMEs and service providers. Service providers provide the actual service to the SME. Business resource brokerage organizations span these two synergistic categories of organization.

5. **The benefits of successful networking must outweigh the costs.**
   Developing formal value added networks is time consuming and costly and must involve a positive sum process to be justified. High growth firms have found the cost-per-benefit to be worthwhile. (Networks of Innovators 1990)

6. **Network brokers are valuable but rare.**
   The networking process can be made less costly and more accessible by networking brokers, responsible for guiding the collaboration process. Networking brokers must be exceptionally competent, technically aware and industry-credible individuals with strong people skills. Such individuals are not easy to find. (Scott, Jones, Bramley, and Bolton 1996)

networks. It is especially important to identify the networking processes that can open the more remote knowledge resources of research institutes and universities to smaller innovative high technology firms. How do we unleash the broadest scope of knowledge in support of innovation? Again, the framework of our interviews tended to focus the discussion on institutional components of the regional innovation system.

Theoretical Framework

The phrase "knowledge economy" has become a mantra. There is substance in it, but the core issue is not simply the value of knowledge. What is central is the changing environment for the acquisition, control, and use of knowledge and the character of the knowledge required. The drivers of the second industrial revolution of the late 19th and early 20th centuries (Landes 1969), were new materials and sources of energy (steel, chemicals, electricity, internal combustion engines). They grew within organizations capable of internal mastery of the evolving knowledge. As well, that knowledge was, more often than today, the tacit knowledge (Nonaka and Takeuchi 1995) of the shop floor rather than the explicit knowledge of the R&D report. Out of the second industrial revolution, industrial research, built on growing scientific disciplines in electricity and chemistry, emerged (Noble 1977) to increase reliance on explicit disciplinary knowledge. (It must be remembered that tacit knowledge remained crucial. The growth of the steel industry provides a clear example.[1])

The current context has evolved considerably from that of the second industrial revolution. Rewarding innovation problems have developed a much more complex character. The situation has been described as what is now widely called the "new production of knowledge" (Gibbons, et al.1994). The new production is characterized by transdisciplinary synthesis of explicit knowledge from scientific disciplines with tacit knowledge arising within firms, markets, and supplier relationships. Few rewarding problems draw on only one discipline. Earlier models of the role of science in innovation relied on a linear concept that envisioned a sequence from disciplinary discovery to invention, to development, to commercialization. This vectorial

[1] Comment on the steel industry presents an opportunity to comment on the distinction between product and process innovation. The innovations in the chemical industry of the second industrial revolution brought forward many new products. Steel was well known before the second industrial revolution and the innovations were in manufacturing process.

"technology transfer" concept has been widely criticized recently (e.g., de la Mothe 1994). As we struggle to characterize the topology of the innovation process with images such as neural nets or chain links, we recognize that problems confronted by innovators depend at least as much on the configuration of existing knowledge as on new discovery. We must not underrate the creative challenge posed by transdisciplinary knowledge configuration.

It has been argued forcefully (Gibbons et al. 1994) that even the largest firms can no longer control internal access to the full range of knowledge needed.

Empirically, this argument seems to be well supported by the strong movements toward identification of core businesses and toward outsourcing. If even the largest firms cannot internalize all knowledge production, it follows that small innovative firms will have even greater needs for external inputs. Thus, the current structure of knowledge production and exploitation creates a compelling need for networking as a tool for knowledge aquisition.

A brief digression is warranted here. A key issue is the translation of the relationship between the explicit knowledge of the research report or patent document and the tacit knowledge of the shop floor. Nonaka and Takeuchi (1995) have analyzed the processes. Tacit knowledge is shared through a socialization process. Apprentices work with masters. The key is experience sharing. Externalization, the process of converting tacit knowledge to explicit, is a process of concept development, usually by metaphor creation and model building. It is often triggered by dialogue and collective reflection. Externalization is essential to generalization of knowledge for an organization. Systematizing explicit knowledge into a knowledge system by appropriate configuration of the existing knowledge is a process called combination. Finally, internalization is the process by which the results of the other three translation processes are converted by an individual into tacit knowledge. Manuals and the opportunity to practice are valuable aids to internalization. In a healthy situation, these four interact in a continuing, reflexive, knowledge spiral.

Much of the literature on networking (see Ebers 1997 for a recent review) focuses on the networking relations that exist among a firm and its suppliers and customers. Such relationships do include a large element of knowledge acquisition and configuration, and are regularly identified by entrepreneurs as major sources of information. Other sources of knowledge receive less attention. Government agencies, research institutes, universities, and even consulting firms figure less prominently in reports on knowledge acquisition. One reason is the translation problem, especially combination and socialization processes. We tend not to appreciate the complexity of translation. Latour (1999) provides a good example of how complicated translation can be, even within a single scientific discipline, using a problem in ecology. Careful observation of a project to study the relative movement of the boundary between the savanna and the rainforest in Brazil elucidates the many steps between fieldwork at the forest boundary and the composition of a scientific paper in Paris. The words in the paper link to the field in a complex circulating reference documented by Latour with 26 illustrations (photos, diagrams) of the steps. In our context, it is easy to recognize that knowledge drawn from sources most closely linked to the institution minimizes the still formidable translation problem through easier socialization, externalization, and combination. The steps in translation between a firm and a supplier are fewer and less complex than, for example, the steps of translation between a firm and a university laboratory. However, the most effective configuration of knowledge requires effective translation to remove the limits from the tools available to the innovative firm. We will return to the critical issue of translation in the penultimate section of this paper.

Methodology

The six propositions about networking practice that formed the basis for the interviews are set out in Table 1.[2] These propositions formed the basis for the series of interviews, but were not used in a limiting way. The interviews were open-ended and encouraged the respondents to discuss their experiences broadly. The propositions provided the starting point. The six tabulated propositions were well received in all interviews, and their points were illustrated by many experiences of the interviewees. As well, much of the detailed structure of networking interactions was discussed. This paper also draws on the experience of the MettNet Institute, a network broker organization that worked with innovative growth firms from 1992 through 2000 to assist them in identifying resources, especially knowledge resources, that would overcome limits on growth.

The interviews are focused on two regions, the Calgary metropolitan area and the Ottawa-Carleton region. It is important to note that the six propositions that fueled this study did not lead to discussion of

[2] For a related set see European Commission (1999).

taxation, research credits, patent regimes, or the sources of outside financing of the research institutions that function outside regional innovation systems. Thus, our interviews illuminated the regional innovation system aspects. They do not provide insight into the importance of the national infrastructure except as national organizations function directly within the regions.

This paper seeks to exploit the comments generated around (rather more than directly about) the six propositions, and the MettNet experience, as the basis for analysis of the subtle processes of knowledge transfer and translation that took place through the networks known to the interviewees. The paper seeks to identify translation tools and strategies used to promote effective knowledge dissemination. Of special interest will be the processes of translation that facilitate knowledge transfer between organizations operating in significantly different cultural milieux. This area of interest reflects the contention that it is the broadest knowledge base that is most conducive to substantial innovation.

Main Elements of, and Conditions for, Networking

Ebers' review (1997) identifies three major elements of network activity: resource flows, mutual expectations, and information flows. Ebers suggests that these three can be operationalized using the terms activity links, trust, and catalysts.

Activity links. Activity links are an aspect of networking engaged in by firms of all sizes and stages of development. Links include subcontracting, outsourcing, and similar phenomena. The economic drivers are well understood. In contrast, the effects of the changes in the production and flow of knowledge are less well analyzed. Our interviews identified reasons for networking in terms of performing tasks that cannot be accomplished alone and developing long-term relationships. Thus, it is evident that activity links have an important dimension of knowledge exchange, much of it tacit. Organizational meeting grounds for firms with similar problems and experience, especially emerging clusters in one industry, can be major facilitators of activity links with simple translation requirements. One of our contributors, Tenet Medical Engineering of Calgary, illustrates how much further activity links can extend into knowledge production. The Medical Devices Development Project (MDDP) exploits a university design course as the primary stage for projects with "no burning timetable." The absence of a "burning timetable" reflects needs of the translation process. Tenet takes one or two projects a year from the MDDP for commercialization.

Trust. If actor-network theory, the leading theoretical scheme in science studies, suggests the highly egocentric view that actors conceive a network as a tool for enlisting other actors IN their interests, the emphasis on the role of trust in networking provides the balancing consideration. Trust is the enabling condition that facilitates the formation of ongoing networks (Luhmann 1979). The overall concept of trust can be divided into three major subcategories (Hartman 2000). These are competence trust (confidence in the capacity of other actors to perform), business trust (confidence in reliability in transactions), and emotional trust (personal confidence based on personal relationships). Distrust of government programs is often expressed in innovative growth firms and reflects a translation problem in business trust, a difference in perception of financial relationships. One of the clearest expressions of competence trust was an interview observation from one firm that a network partner representative "can address [critical issues] in a way that is highly effective but does not take up a great deal of the firm's scarce resource or the time of key personnel," inferring that this networking representative's work does not require checking!

Catalysts. This element is the central concern of the present paper. Catalysts, or informational intermediaries, may be people or inter-organizational information systems. Their role facilitates new modes of knowledge production. A European research network that began with a conventional set of market-defined relationships evolved from a "zero sum game" into a "positive sum game," as Lutz (1997) notes, when actors were stimulated to give up well known rules of procedure and to explore alternative problem-solving strategies. Two research institutes played catalytic roles. The job of the catalyst was to support translation to ways of thinking about rules of procedure that opened the transdisciplinary pathways to much more substantial innovation. Notice, as Lutz's case study demonstrates, that the process is evolutionary, not static.

We will consider three types of actors that explicitly undertake translation roles (catalysts). The first type is the community agency oriented toward economic development. The examples here will be the Calgary Research and Development Authority (CRDA) and the Ottawa Centre for Research and Innovation (OCRI). The second type is a regional consortium of firms. Examples include the Ottawa Carleton Manufacturer's Network (OCMN), the Calgary Council for Advanced Technology (CCAT), and the Calgary Association for Medical Products (CAMP). The third type is the explicit interventionists and brokers oriented toward facilitation of innovation. This type includes both broad inter-organizational systems such as the Canadian Technology Network (CTN) and more narrowly focused actors that focus exclusively on assistance to individual firms, including finding network partners. Examples of the latter include the National Research Councilís Industrial Research Assistance Program (IRAP); Ottawaís DYI Technologies, Inc., and the MettNet Institute. Notice that these catalytic agencies include mainly actors involved in the construction of regional innovation systems.

A final note: several of the catalysts we consider can also be designated institutions. Edquist (1997) has made the point that institutions are of crucial importance for the innovative process. The innovation systems approach places "institutions" in a central role in all variants and that is a strength of the "systems of innovation" idea. However, institutions will not explicitly reappear in the discussion of pathways of knowledge until we reach the end of our analysis, despite the fact that our discussion of the components of regional systems implicitly considers their institutional roles.

The Role of Community Development Agencies, CRDA and OCRI

The Calgary Research and Development Authority (CRDA, now known as Calgary Technologies, Inc.) is a part of the City of Calgary's promotional structure. A consortium comprised of the City, the University of Calgary, and the Calgary Chamber of Commerce owns CRDA. It was founded at the instigation of oil industry leaders concerned with the diversification of the Calgary economy. The CRDA operates as a promotional agency to encourage high tech firms to locate in Calgary, and operates a research park.

As a catalyst it operates at two levels. At one level, CRDA brings together senior representatives of large high tech players, maturing smaller firms, and public sector research organizations. Groups discuss issues affecting the Calgary high tech sector. Participants include representatives of major high tech firms (e.g., Nortel), creative younger firms (e.g., Smart Technologies), and technology clusters such as the geographical positioning systems (GPS) sector, along with university and hospital research actors. The discussion sessions and shared visioning are effective trust builders. They support development of both competence and emotional trust. From that base, the group can develop strong cross-institutional consensus positions that influence policy. The representatives of the various sectors translate the language of the others in terms that can be internalized into their own institutional perspectives. One consequence has been effective articulation of the University of Calgary's role in the community of informing government planners. Consultations also established a political base for promoting enhanced human resource strategies for information technology.

At another level, the CRDA is the manager of an incubator for young firms and hosts an Innovation Centre sponsored by the National Research Council (NRC) that seeks to identify opportunities for firm formation, especially in the health area. The incubator primarily facilitates sharing experience among the participating firms, but also provides access to business expertise (explicit knowledge). Information transfers are of the sort that arise in consortia of firms and do not involve the most difficult translations. This kind of networking is further discussed in the next section, the role of local consortia. The Innovation Centre undertakes the higher-risk effort of bringing together knowledge from diverse sources, including academic and hospital environments, to facilitate formation of new firms. The processes required are discussed later in the role of inventors and brokers section.

The Ottawa Centre for Research and Innovation (OCRI) is, in many respects, similar to the CRDA except that it is a larger organization and offers a wider spectrum of media for the communication of explicit knowledge. OCRI seminars include technological, legal, and business topics. Its linkages encourage access to facilities developed by the federal government, and regional educational institutions are actively involved. The OCRI is an excellent example of an organization concerned with transdisciplinary knowledge production and dissemination that starts with the combination of explicit knowledge. A superficial examination of the activities that are explicitly organized by OCRI (in contrast to the facilitation it provides for access to government and university resources) suggests that OCRI is primarily involved in networks for knowledge combination. Indeed, OCRI President Bill Collins commented to our interviewer:

> At OCRI events, people come together to learn something or do something relevant to a project of their own. Companies that are successful in Ottawa are cautious networkers. They go to events that fit them and they don't get caught up with networking only as a social event. They assign different staff to different events.

These are characteristics of explicit knowledge combination.

There is a local adage in Ottawa: "More technology transfer happens in a bar than ever goes on in a lab." This remark underlines the fact that

OCRI networking also builds trust, at the very least on the emotional level. Trust is an important precursor for the development of the socialization processes for tacit knowledge transfer. Collins also noted to us: "Networking builds trust. This works in Ottawa where there is a family structure, at some point they all worked in the same company, so trust can be built earlier." In other words, a shared core of common language facilitates translation. This is the environment for development of active partnering relationships.

It is important not to overlook OCRIís impact on more complex translation. It has been an active promoter of projects of such as university/industry collaborative research chairs. These chairs promote an ongoing relationship involving flexible joint research programs that can adapt to the timetables and cultures of both academia and industry. Their flexibility facilitates knowledge flow (both ways) in a fashion that allows each institution to perceive itself as an integrating system that can express the knowledge in its own working language.

The Role of Local Consortia

Both OCRI and CRDA include the facilitation of peer communication among comparable firms as a part of their activity. There are also free-standing organizations that play this role. In Calgary, the Calgary Council for Advanced Technology (CCAT) and the Calgary Association for Medical Products (CAMP) are examples. In Ottawa, the Ottawa Carleton Manufacturerís Network (OCMN) is an example.

The CCAT organizes regular meetings for presentations relevant to young firms in technology-rich industries.[3] The meeting format is oriented toward explicit knowledge presentations. However, it is clear that some translation of tacit to explicit knowledge occurs in the context of these meetings. Gary Bishop of Jambalaya characterized how this arises to our interviewer: "Networking with companies that have gone through growth and know the pitfalls to avoid is of great use to smaller companies." This leads to activity links. Regional clusters in the same industry can foster activity links readily, as illustrated by the other Calgary consortium, CAMP. This is an association of the growing group of Calgary companies in the medical products field. Ken Moore, President of Tenet and a founder of CAMP, described the rationale.

[3] The firms networked through CCAT are typically at an earlier stage than the ones that participate in CRDA meetings. They are closer to the CRDA incubator participants.

> I am involved in CAMP as I think we must locally grow our capacity in medical products. The first step is getting to know who are the players so that we can network, meet and identify our needs and forge partnerships where relevant. (We have had companies realize that there is a needed product supplier locally) People are teaming up to develop new products and to jointly beach-head new markets.

This illustrates a consortium serving as the channel for the connections that lead to both strategic partnerships with suppliers and strategic joint ventures. The prime functions of the consortium are probably in the area of trust building, including the three subsections of competence, business, and emotional trust. Trust permits the partners to explore the kinds of close relationships (socialization) in which tacit knowledge transfers occur along with the positive sum game of direct economic benefits from strategic relationships. It is likely that similar relationships do arise from the looser associations, as with CCAT, or the circles that arise within OCRI or the CRDA incubator. However, the network created by a regional industrial cluster has obvious advantages.

A consortium that lies between CAMP and OCRI in the degree to which the member firms share common interests is the Ottawa-based OCMN. The network has member companies and individuals who share the goal of enhancing manufacturing competitiveness and productivity and who want to share knowledge. The menu of regular OCMN activities includes round table and case discussions, small group meetings on mutually chosen topics, guest speakers, short courses, and plant visits. There is an obvious mix of opportunities for different modes of knowledge transfer. A list of OCMN accomplishments includes the following items:

- borrowing practices implemented,
- joint software development projects completed,
- resource sharing for cost reduction,
- collaborative projects,
- access to university-based research and students.

The list is impressive and testifies to effective catalysis, including some subtle translation from the academic to the business domains. Much of the success is attributed to the work of the coordinator, who appears to function in the role of network broker as discussed in the next section.

The Role of Intervenors and Brokers

There has been one actor that has intervened at the national level in Canada specifically to promote knowledge exchange and translation in support of innovation by smaller firms. This is the ambitious Canadian Technology Network (CTN) sponsored by the NRC (and closely related to IRAP). The CTN strategy is to draw on most research agencies, trade associations, universities, government laboratories, and related bodies in the country as advisory or affiliate members. The web site listing of these members with the brief accounts of their capacities is a significant resource in itself. Advisory members are institutions that provide the services of a network advisor who will respond as a Subject Matter Expert (SME) to inquire and work through the structure of the situation to identify needs of, and solutions to, technology challenges. It is recognized that technology cannot be separated from related business assistance. The advisor will use local knowledge and the information available through the national network of the CTN to find the sources of support needed by the SME. In many cases, these sources are found in the affiliate members who are open to establishing working relationships with outside clients.

The most frequently sought forms of assistance are:

- overall management (HR, training policies, regulation),
- marketing management (development of markets, exporting),
- technology management (evaluation, lab testing, development, IP),
- financial and business planning.

CTN's ambition is to achieve the widest scope of SME networking with knowledge sources for both explicit and tacit knowledge acquisition. The success of the advisors depends on the skills they develop in the translation process for linking SMEs with large corporate partners, financial institutions, research institutes, and other sources. Since each advisor has a connection with a particular advisory member institution, the challenge faced by the advisors is considerable. They must learn the culture of a wide variety of institutions and develop next level translation skills. As might be expected, the outcomes depend on the details of the client advisor relationship and the openness of affiliates to provision of ready access. In large organizations, affiliate contact points may have serious difficulty making contact with the appropriate internal resources. The scope to which the CTN plan aspires is in itself a limitation. It is a bit daunting to deal with such a comprehensive list of resources. In practice, the national organization may more often function with local actors as participants in the regional innovation systems that have figured in most of this account. CTN then resembles its institutional relative, IRAP, in important ways.

The NRC Industrial Research Assistance Program (IRAP) is widely recognized as one of the most successful government programs supporting technological innovation. Of course, this is a function of its success in identifying good projects to finance, since it can provide material support. The program is delivered through a network of industrial technology advisors (ITAs). Their function similar to that of CTN advisors, and many (though not all) IRAP-ITAs are designated as CTN advisors. An early responsibility of ITAs was to find ways that NRC could provide technical support. In the case of growth SMEs, this requires development of skills in translation between the knowledge spheres of a smaller company and a large government research establishment. Such discussions usually begin around the explicit knowledge that the laboratory can bring to bear on the company's concerns. However, the discussions can easily break down if the ITA cannot develop translation mechanisms that link the tacit knowledge of the laboratory to the tacit knowledge of the firm and dynamically connect the cultures of the two so that each recognizes solutions in the language of its work.

IRAP's role as a source of matching fund financial support to innovative technology projects in SMEs also affects knowledge transfer. We have learned (through some firms with whom MettNet has worked) that this feature can pose a barrier to development of business trust relationships that are important to an ITA as a networking agent. It is the obvious analog to the dilemma faced by teachers, the task of being at the same time a mentor and an evaluator.

The ITA networks extend well beyond NRC laboratories. Individual ITAs are active in exploiting a variety of knowledge sources, especially provincial research agencies and universities, for technology assistance. ITAs often work on business plans and have also been key actors in identifying pathways for the commercialization of technologies emerging from research institutions such as universities.

The CTN and IRAP are large and complex networking organizations devoted to knowledge brokering. Their governmental mandates and the scale and scope of their activities may limit the degree of intimacy of relationships with individual SMEs. The last two actors to be described are smaller, more focused, independent networking and knowledge brokers, DYI and MettNet. DYI Technologies, Inc. is an Ottawa consulting firm that works to build business networks and consults in various areas of advanced technology and organizational development. The Principal of the firm, Andrew Young, has experience with large firms and is an advisor to Inco Mines Research. He also has academic connections in the areas of operations management and computer-integrated manufacturing. He articulates the role of translation very clearly. His goal is to take ideas from different domains and apply them to new domains. In our interview, he characterized innovation informally as "looking from different points of view and asking silly questions." He was a founder of the OCMN, and his translation approach is credited with important parts of the success of that network.

The MettNet Institute operated from August 1992 through March 2000. It was initially promoted by a group of academic and industrial leaders associated with metallurgical industries, but the breadth of its mandate grew rapidly. Its sponsors included universities, banks, large firms, and the Natural Sciences and Engineering Research Council (NSERC), but growth SMEs have been the client focus. It operated in Ontario, Quebec, Alberta, and Nova Scotia. In each case, local networking was important and MettNet became an actor in a regional innovation system. However, MettNet was always sensitive to resources identified in its national activity. MettNet demonstrated in several cases that regional proximity is not essential to close network relationships.

MettNet developed an outreach model for involvement with SMEs that has five steps:

- Proactive identification of technology-driven, high growth companies.
- Client company-driven, non-partisan, global needs analysis.
- Networking with knowledge and service providers for solutions.
- Follow-up to validate action.
- Ongoing contact with company and networking partners to "project manage" relationships and respond to evolving needs.

Note that companies are sought out, but criteria for inclusion are applied. The model represents an innovation in technology networking because technical, financial, and management needs are regarded as indissociable. This dictates a global, transdisciplinary approach to identifying company needs and determines the choice of networking partners. The approach is emphatically client-driven. By beginning with the needs elicited from the company and networking backward from needs to appropriate resources, the outreach effort evolved to encompass resources from all of technical, management, and financial resources. The broad range of resources required translation among the cultures of the SMEs and, for example, universities and large banks. The complex circulating reference (Latour 1999) of the dynamic and reflexive translation process requires ongoing contact and project management by the network broker. The value of continuing project management cannot be overemphasized.

MettNet has been particularly associated with universities as resources. In this regard, it is interesting that one of the frequent outcomes of MettNet interaction has been placement of appropriate recent graduates or co-op students as tools for enriching the technological capacity of a company. This reflects that the migration of persons is one of the most effective methods of transmitting tacit knowledge. Another interesting result of the MettNet experience is that university discovery research projects are less frequently the pathway of productive networking than interactions that assist the firm in the configuration of existing knowledge, a key aspect of the new production of knowledge (Gibbons et al 1994). An important point to recognize is that the translation process acting over the whole knowledge spiral of Nonaka and Takeuchi (1995) can produce the transdisciplinary and dynamic aspect that makes the configuration of knowledge process into knowledge production.

Translation and pathways of knowledge

In our interview, Gary Bishop of Jambalaya Technologies, Inc. summarized a view from within of the networking process.

Most people network in their comfort zone, and this does not bring innovation. Do CEOs know when they have a hole in their armour and if their networks are sufficiently large that they can solve your problems? You need to network outside your comfort zone to learn more and to know what you don't know. To address your gaps, you have to look at different networks. This is the true value a network broker can bring, extending a company's network into new areas.

The summary of networking activity presented in this paper suggests a division of the translations on the pathways of knowledge into two levels (Leydesdorff 1997). The first level is most common to networking between firms with recognizable related interests. It often occurs among firms in a network generated either by a peer group association or a regional agency. This is, perhaps, mainly networking in the comfort zone. The relationships between firms can be diagrammed as shown in Figure 1. Here three firms (a, b, and c) have overlapping interests represented by the area i. The area i may be thought of as representing a common language, a discursive system following a single logic. There is differentiation of communication, but a common zone exists. All of the specialized jargons of the firms, including features of tacit knowledge, can be translated in terms of the language of i, if they can be sufficiently elaborated using processes discussed by Nonaka and Takeuchi (1995). Most of the activities we have described belong to this level. They support innovation to the extent that activity links are facilitated and learning allows for avoidance of pitfalls. These translations contribute to an intellectual hygiene of the firm, essential to the success of its innovations, but they are probably not the translations that initiate and inspire profound innovations.

Leydesdorff (1997, 112) suggests that our contemporary translation systems "are no longer discursive systems that follow a single logic, but next order translations that recursively entrain the various logics that go into them as their 'genotypes' The new [systems are] based on a multitude of perspectives." Consider Figure 2. Here the three participants (labeled f, g, and u) have differentiated sufficiently that there is no longer an overlapping centre. The common zone of interpenetration has disappeared. The central integrating circle is not expected to exist at any single moment. The system is integrated over time. Messages are passed or received sequentially and recursively, implying translation at a receiving end. Each of f, g, and u can perceive itself as the integrating system since each has its own criteria for integration.

Let's imagine that the systems f, g, and u represent firms, government agencies, and universities. It is our contention that this second order dynamic translation process must be addressed in integrating knowledge over firms, government agencies, and universities. The time dependence and the reflexivity create the requirement for continuing relationships and project management on the part of catalytic agencies. The reflexive passing of messages among the sectors creates an evolutionary trajectory. Clearly, it is a difficult and delicate process to select particular trajectories over time. The subtle goal is stabilization of the systems into co-evolution.

Figure 1. Three firms; a,b, and c, communicate via translation of messages indicated by the common language area I shared by all.

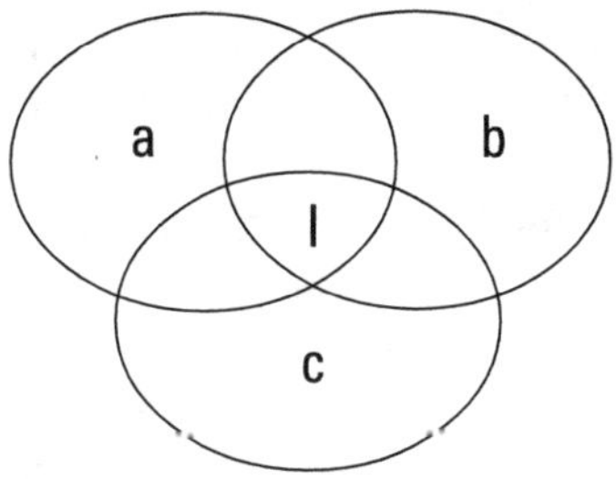

Figure 2. Three participants communicate in a dynamic reflexive process over time. There is no central area of common language and each will see itself as the integrating system since it has its own criteria for integration.

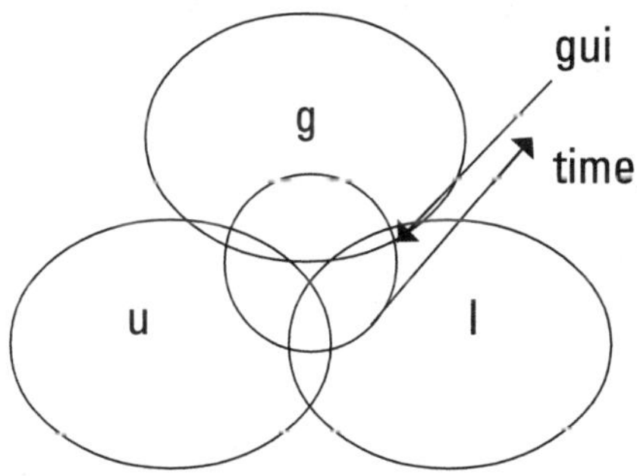

Notes on National, Regional, and Interregional Innovation Systems and on University Roles

This analysis of catalytic agencies and their role in knowledge flows offers a perspective on the relationships of various levels of innovation systems. The activities identified in our interviews tend to have a local focus and emphasize the benefits of regional innovation system analysis. The MettNet experience shows, nevertheless, that non-local networking can be fruitful.

Finally, an opportunity for universities can be identified. If a second order reflexive dynamic translation process must be addressed, the language widely used to discuss university technology transfer has become insufficient and, perhaps, misleading (Leydesdorff 1997). For example, Kaghan and Barnett (1997) have articulated and illustrated the argument that the language of technology transfer overemphasizes the laboratory model of university contribution and does not appropriately allow for a desktop model, where contributions from universities are in a creative knowledge configuration. Similarly, emphasis needs to be given not to single-event university-industry interactions, but to ongoing relationships that can evolve and allow both the industry and university receptors to exploit their own criteria for integration within an evolving relationship. In practical terms, the most effective model for exploitation of the university may be the industrial research chair, where a multidimensional relationship with a commitment to a degree of continuity permits fruitful co-evolution. In the technology transfer office (better called an industry liaison office), more emphasis needs to be given to developing network brokers who can project manage individual relationships with the long-term attention to detail characteristic of organizations such as DYI Technologies, Inc. and MettNet. Unfortunately, proposition six in Table 1 reminds us that network brokers are rare, a conclusion shared by a recent European study (European Commission 1999).

Acknowledgments

We are much indebted to the six executives who gave a considerable amount of their valuable time to inform us about their experiences with networking in relation to innovation. We hope they may still recognize some of their thinking in the interpretations that we present, because their insights proved the key stimulus to our thinking. Clearly, the interpretations are our responsibility. Sincere thanks to:

Ken Moore, President, Tenet Medical Engineering, Inc.
Al Leonard, President, WorkPlace Wisdom, Inc.
Mike Horsfall, GIS Specialist, WorkPlace Wisdom, Inc.
Gary Bishop, Vice-president, Jambalaya Technologies, Inc.
Andrew Young, Principal, DYI Technologies, Inc.
Bill Collins, President, Ottawa Centre for Research and Innovation.

References

de la Mothe, J. (1994). Canada and the national innovation system. In, Resource Book for Science and Technology Policy Review Consultations, vol. II, 11-20. Ottawa: Industry Canada.

Ebers, M. (1997). The Formation of Inter-Organizational Networks. Oxford: Oxford University Press.

Edquist, C. (1997). Technologies, institutions, and organizations. In Systems of Innovation: Institutions and Organizations, ed. C. Edquist. London: Pinter.

European Commission Studies, Science, Research, Development (1999), The Globalizing Learning Economy: Implications for Innovation Policy. Luxembourg: Office for Official Publications of the European Commission, EUR 18307 EN.

Gibbons, M., Limoges, C., Nowotny, H., Schwartzman, S., Scott, P., and Trow, M. (1994). The New Production of Knowledge. London: Sage.

Hartman, F. (2000). Don't Park Your Brain Outside. Calgary: Project Management Institute.

Kaghan, W.N. and Barnett, B. (1997). The desktop model of innovation in digital media. In Universities and the Global Knowledge Economy, ed. H. Etkowicz and L. Leydesdorff. London: Pinter.

Landes, D. (1969). The Unbound Prometheus. Cambridge: Cambridge University Press.

Latour, B. (1999). Pandora's Hope: The Reality of Science Studies. Cambridge: Harvard University Press.

Leydesdorff, L. (1997). The new communications regime of university-industry-government relations. In Universities and the Global Knowledge Economy. ed. H. Etkowicz and L. Leydesdorff. London: Pinter.

Luhmann, N. (1979). Trust and Power. Chichester: John Wiley and Sons.

Lutz, S. (1997). Learning through intermediaries: The case of inter-firm research collaborations. In The Formation of Inter-Organizational Networks, ed. M. Ebers. Oxford: Oxford University Press.

Networks of Innovators (1990). Montreal workshop. Research Policy, 1991, 20, nr5.

Noble, D. (1977). America by Design. New York: Knopf.

Nonaka, I. and Takeuchi, H. (1995). The Knowledge-Creating Company. Oxford: Oxford University Press.

Polanyi, M. (1958), Personal Knowledge. Chicago: The University of Chicago Press.

Steier, L., and Greenwood, R. (1999). Newly created firms and informal angel investors: a four-stage model of network development. Venture Capital 1(2): 147.

Scott, P., Jones, B., Bramley, A., and Bolton, B. (1996). Networking brokers. International Small Business Journal 14: 85-99.

III

Innovations in Entrepreneurial Support Programs

Chapter 9

Lessons from Successful Innovations in Entrepreneurial Support Programming

W. Ed McMullan, James J. Chrisman, and Karl H. Vesper

Abstract

The value of entrepreneurial firms in the economic development of a region or nation is well known. Over the past two decades, several successful programs designed to assist entrepreneurial firms have been launched in western Canada, the United States, and Australia. Despite these successes, it is our contention that these successes have not been exploited as fully as they could and should have been in western Canada. In this paper we will describe some of the success stories in entrepreneurial support and provide some advice for the design and evaluation of future entrepreneurial assistance programs in Western Canada.

Introduction

David Birch's seminal initial (1979) and follow-up (1987) work on job creation focused the attention of public policy makers on the economic contributions of small, new firms. The flurry of interest in Birch's work was accompanied by a host of new programs designed to facilitate the development of small companies.[1] While there was a lot of interest, little formal knowledge existed on how to economically assist

[1] Entrepreneurial support programming includes any programmatic expenditure to help entrepreneurs develop businesses. This involves entrepreneurs in pre-start, start-up, and ongoing business situations. The assistance can consist of anything from education to work space subsidization to venture funding. Education programs in particular can vary dramatically, from 13-week courses to Master's Degree programs.

entrepreneurs with their business ventures. Although much was learned in the 1980s and 1990s, there are still relatively few economic development programs that have been thoroughly vetted. Evaluations of programs designed to assist or educate entrepreneurs and small business managers are important because such programs are typically subsidized by public money or private grants, and therefore are not subject to the discipline of the marketplace. Without program evaluations it is difficult to determine if funds expended on programs designed to assist entrepreneurs are worthwhile investments of society's scarce resources.

Since it is typically not in the interests of either public financiers or program administrators to take responsibility for wasting money on bad designs, it is not surprising that many entrepreneurial support programs are not properly evaluated (Bartik 1994). Our experience suggests that only a small portion of the entrepreneurial support programs in existence have been subjected to good evaluation methodology. In an era of tax cutting and calls for government accountability, it becomes increasingly important for the public sector to properly justify public expenditures.

Existing program evaluations allow us to identify designs with significant promise of producing an economic impact. In this article we shall discuss several different programs that have worked in the U.S., Australia, and western Canada. We will point to evidence of their effectiveness and discuss the ingredients of the programs that we believe have contributed to their effectiveness. Our purpose is to identify program designs that are worthy of being adopted, adapted, and expanded in western Canada and elsewhere as part of good economic development strategies. A secondary purpose is to emphasize the importance of objective evaluations, both to ensure that entrepreneurial assistance programs are adequately funded and that their economic development objectives are achieved.

A Review of Entrepreneurship Assistance Programs

In this section we will review several entrepreneurial assistance programs with which we have extensive familiarity. Appendix 1 provides summary descriptions of each of these programs. It is noteworthy, we believe, that programs developed outside Canada have not yet been adopted in this country. Empirical support has been generated for the effectiveness of each of these programs. In our opinion, these programs each merit serious consideration for adoption in western Canada. Overlooking such programs is the equivalent of missing good opportunities for economic development.

Small Business Development Center Program

The Small Business Develop Center (SBDC) program was created in the U.S. in the late 1970s by William Flewellen from the University of Georgia. By the mid-1990s it was present in all 50 states, Washington D.C., and three U.S. territories. The program was primarily designed to provide one-on-one counseling assistance to pre-start and ongoing businesses. The program currently costs the American taxpayer approximately $180 million per year, as the services are usually provided free to all comers.

The counseling process involves a series of interactions between the counselor and the client, and specific work assignments carried out by both parties. In many ways the process resembles a directed or independent study wherein the client performs the majority of the work under the guidance and direction of a trained professional counselor. The counselor's time is thereby leveraged, allowing a greater number of clients to be assisted than would be the case if the counselor performed all the work. Furthermore, because the clients are responsible for the quality of their own work, a great deal of learning takes place that can last far beyond the counseling period (Chrisman 1999).

The first publication of a formal evaluation of the program was in 1985 by Chrisman, Nelson, Hoy, and Robinson. That study has been followed by a number of subsequent evaluation studies using essentially the same methodology (e.g., Chrisman, Hoy and Robinson 1987; Chrisman and Katrishen 1994). Congress has repeatedly challenged the viability of the SBDC program, but despite a lack of support from any interest group, the program has withstood their scrutiny year after year. One reason is probably the satisfaction of participants; however, the major reason that the SBDC program continues to thrive is that the data have, time and again, shown a direct contribution to business formation and development.

Evaluations of the SBDC program have focused exclusively on the impact of "long-term" clients, defined as those who receive five or more hours of counseling assistance. The average number of hours of counseling assistance per client is between 15 and 20 hours at a cost of approximately $2000 per client. The foremost question for evaluators at the current time is whether the investments of time and money are adequately rewarded with results. On a national basis the benefit to cost ratios have ranged from 2/1 to 4.5/1, where benefits are calculated from net sales and employment taxes generated. Taxes are calculated from client increases in sales and employment after a one-year time lapse.

They are then compared to the full cost tax dollars spent on the *entire* program (e.g., Chrisman and Katrishen 1994). The only benefits that are counted are those over and above the average growth rates in each state involved.

There has been an ongoing academic debate about the methodology and conceptual foundations of these impact evaluations (Elstrott 1987; Chrisman, Hoy, Robinson and Nelson 1987; Wood 1994; Chrisman and McMullan 1996; Wood 1999; Chrisman and McMullan 2002). On balance, however, the evidence still points to a positive net economic contribution.[2] Furthermore, a recent follow-up study was conducted on firms three to five years beyond the SBDC intervention (Chrisman and McMullan, 2000). That study suggests that (1) the SBDC program reaches new ventures with high potential, (2) the long-term performance of clients with respected to survival, sales growth, employment growth, and innovation is high, and (3) the clients continue to attribute a significant portion of their success to the interventions of the SBDC.

Fastrac/Western Entrepreneurial Network

Fastrac was developed by Dr. Courtney Price from Metropolitan University in Denver in the mid-1980s. The Western Entrepreneurial Network (WEN) was a mid-1990s spin-off from Fastrac. Both organizations are rushing to spread their programming abroad and throughout every state in the U.S.

Fastrac/WEN are typically 13-week course-based programs consisting of lectures, guest speakers, group interactions, and business plan development. Separate courses are available for both pre-start and existing businesses. Each program is subsidized by a different charitable body. Fastrac is subsidized by the Kaufman Foundation and WEN is subsidized by the U.S. West Foundation. The programs typically cost approximately $500 per client, of which the client pays about a third.

In 1998 we conducted an evaluation of the effectiveness of the WEN program (Chrisman, McMullan, and Vesper 1998). Our findings demonstrate positive program benefits in the areas of: preparation and knowledge acquisition, instilling confidence in the entrepreneur, and objective measures of economic impact. These findings were consistent with those of earlier work on the Fastrac program (Monroe, Price and

[2] Because there are numerous assumptions implicit in the findings, the reader should review the original studies and the ongoing dialogues in greater depth.

Burns undated; Monroe, Allen, and Price 1995; Monroe, Price, Allen, and Monroe undated; Price and Prinster 1996; Price, Burns, and Monroe undated). Although the results of the Fastrac and WEN programs do not appear to be as impressive as those generated by the SBDC program, their per client costs are only one-quarter those of the SBDC. Whether Fastrac/WEN clients received more than one-quarter of the benefit of SBDC clients is hard to ascertain because of different screening methodologies used by the programs.[3] However, it appears that the Fastrac/WEN programs are particularly well suited for micro businesses, whereas the SBDC program is more effective in dealing with ventures that are larger in size and have presumably greater growth potential.

Swinburne Master of Entrepreneurship and Innovation

In the late 1980s, Swinburne University in Melbourne, Australia developed a Master of Entrepreneurship and Innovation (MEI) degree program modeled on the work of Drs. Ed McMullan and Wayne Long (1987). Under the direction of Dr. Murray Gillin (now retired), the program produced some noteworthy results (McMullan and Gillin 1998). The program design was apparently unique for its time and does not appear to have been superseded by any program anywhere since.

The Swinburne model was set up as a three-year, half-time program allowing students multiple exit options for the purpose of pursuing their ventures. Successful completion of the first four courses in the program resulted in a Graduate Certificate, the next four in a Graduate Diploma,

[3] One intriguing possibility is that FastTrac essentially produces a one-time jump in performance, while the long-term counseling of the SBDC produces a somewhat longer-term impact as indicated above. Courtney Price and her colleagues conducted three back-to-back studies in the early 1990s on the women and minority graduates of FastTrac, presumably with some overlap among respondents. In the first study, Price and Fleming (undated) found that minority and women FastTrac graduates (small business owners) increased their business performance significantly (e.g., one-year revenue growth of 35%). In a second study of minorities and women, Price, Monroe, and Larsen (1992) found that the program's clients increased employment by 37.2% and sales revenue by 33.5% (both presumably over a two-year period). Again in a third study focusing on women and minorities, Price and Monroe (1993) found, over a three-year period, that full-time employment increased by 27% and sales revenues increased by 33%. This kind of finding demonstrates the requirement for multi-year studies of results.

and the final four in a Master's Degree.[4] Kevin Hindle (personal communication August 2000), head of entrepreneurship research at Swinburne, conceptualized the joint purpose of the different program segments as follows:

1. For the Graduate Certificate: The first four units of the program are intended to provide students with the ability to kill an idea that does not represent a genuine entrepreneurial opportunity. The key concept here is systematic screening. The capstone course of this quartet is Opportunity Evaluation Techniques. The key output is the ability to screen opportunities.
2. For the Graduate Diploma: The purpose of the next four units of the program is to give students the ability to obtain from a variety of stakeholders (who must be persuaded by good evidence and good argument) the necessary resources for a genuine entrepreneurial opportunity. The capstone course of this quartet is The Entrepreneurial Business Plan. The key output is the ability to produce a business plan capable of attracting the capital required to execute a feasible venture plan from an international venture capital firm or appropriate provider of capital. The focus on the business plan also serves a major pedagogical purpose: to stress the multi-disciplinary and interrelated nature of skills required for the execution of successful entrepreneurship rather than treating functional areas such as marketing, finance, accounting, organization behavior, and strategy as virtually separate silos.
3. For the Master's Degree: In the last four units, there are two essences. First, the skill emphasis continues and deepens in intensity and breadth, but so does something else: the ability to think critically and philosophically about entrepreneurship as a social and human phenomenon. The twin concepts in this third and final year are the ability to handle problems in complex, interrelated array rather than one

[4] Students might qualify for the Certificate and Diploma parts of the program but not for the Degree component, although performance in Certificate and Diploma would be counted toward admission in the Master's Degree program.

> at a time, and the ability to understand entrepreneurship in the academic sense (what it has, as a discipline, to contribute to the conversation of humanity). All four courses are capstones in this quartet, but the "primus inter pares" is the Entrepreneurial Research Project.

The program demonstrated an ability to develop applicants into technological entrepreneurs. Although the overall cost of the program was approximately $18,000 per student, the tuition fees covered the entire cost.[5] Eighty-seven percent of graduates started businesses, all of which were reported to have been based on technological innovations. Of those companies that had employees at the time of the survey, the average was eight employees one year out from graduation.[6]

Student Project Assistance Programs

The Small Business Institute (SBI) program that, until recently, operated in the U.S., generated impressive results but was cancelled by the Small Business Administration for reasons unknown.[7] The SBI program involved student teams working with independent businesses in a consulting capacity on a variety of problems, most frequently developing business plans and market research. Studies over a period of years indicated that SBI clients outperform comparable small businesses after receiving assistance (Solomon and Weaver 1983). Overall, clients viewed the assistance received in a favorable light (Weinstein, Nicholls, and

[5] For legal reasons it is unlikely that the program could be replicated in Western Canada without either a public subsidy or government approval to charge students full tuition. In fact, McMullan and Long tried unsuccessfully to implement a Swinburne-style program at the University of Calgary. Dr. Jim Graham was able to introduce a hybrid design intended to develop both entrepreneurs and managers. The Calgary program has, however, produced markedly fewer and less interesting entrepreneurial ventures than the Swinburne design (McMullan and Gough 1999; also see chapter 10 of this book). Here we have an innovation that was implemented successfully in Australia but could not be implemented in western Canada.

[6] Certificate, Diploma and Degree graduates all appear to have done well, based on the data collected to date. Longer-term studies may be required to distinguish real differences in performance by level of academic accomplishment.

[7] The University of Calgary runs a similar outreach program to assist community entrepreneurs using masters and undergraduate business students. It has also generated impressive results (Chrisman, 1997; McMullan, Long and Graham, 1986).

Seaton 1992) and gained significantly different perspectives on their problems as a result of the interventions (Jackson, Vozikis, and Babakus 1992). Furthermore, students who engaged in SBI consulting activities as part of their coursework in business policy appeared to gain more from the experience in terms of analytical and interpersonal skill development and overall satisfaction than students who took a more traditionally designed policy course (Watts and Jackson 1995).[8]

Another type of student project program that has been operated with great success is the Small Business Legal Clinic. Originating at the University of Calgary in the mid-1980s, the program expanded to a number of other university sites in western Canada in the 1990s. This program offers pre-start businesses nearly free legal information from students studying commercial law (Brown, Colborne, and McMullan 1988). In a way, this program is parallel to the student Legal Aid program that is offered to assist the poor. Chrisman's (1997) study of the University of Calgary's Venture Development Program suggests that the Legal Clinic was very successful in meeting client expectations and adding value. In addition, he found that clients used a high percentage of the information garnered from the clinic.

Discussion and Generalizations

Entrepreneurial support can be done well or it can be done badly. Substantial knowledge has been accumulated, and it now makes sense to professionalize entrepreneurial assistance by having economic development professionals choose more results-based programming, and for those seeking to design programs to assimilate existing knowledge as a first step. Evidence has been accumulating on the elements of good and bad design. No longer can program designers, public policy makers, or private funders claim to be acting in a vacuum. More experimentation is definitely needed, but each new experiment needs to be evaluated with an appreciation of the research evidence on program effectiveness.

Program Evaluation

When entrepreneurial support is done well, it can make economic sense. Public accountability means treating entrepreneurship education and

[8] Interestingly, the SBI students did not appear to gain more entrepreneurial aspirations than those involved in a more traditional business policy course. Watts and Jackson (1995) speculated that this may have been due to the former group's greater appreciation, and hence, apprehension, of the difficulties involved in running a small business.

entrepreneurial support systems as social engineering and measuring their effectiveness by the economic growth and development generated (McMullan 1988; McMullan and Vesper 1987). Program performance evaluation is now a must-have component in any serious entrepreneurial support program.

Program evaluation must key on subsequent venture performance, not participant satisfaction. McMullan, Chrisman and Vesper (2001) found that subjective attributions of program value cluster together, and that objective measures of subsequent venture performance cluster together but that these two groups of measures tend to be unrelated. The most credible explanation is that these two types of measures are measuring different domains; the former is likely measuring satisfaction with instruction, while the latter is likely measuring the economic impact of the assistance. Another possible explanation is that certain types of subjective measures, such as perceived program benefits, may be capturing what clients expect to get out of the program over time as they apply the knowledge gained to business situations. For example, Chrisman and Leslie (1989) found that operating and administrative assistance had a significant positive impact on short-term return on sales (ROS) for small business clients, whereas strategic assistance had no such impact. On the other hand, comprehensive assistance that included attention to all three areas had the largest positive influence on ROS. However, in a related study using a similar sample of clients, Chrisman and Carsrud (1989) found that both pre-venture and small business clients value strategic assistance more than either operating or administrative assistance. Apparently, client perceptions of what generates benefit are not always manifested immediately in the bottom line.

The main points are that to properly evaluate programs, we (1) need to be evaluating program impacts and economic outcomes, and (2) need to utilize a variety of measurements. Over the last two decades, the methodology backing outcome evaluation has been developed sufficiently to justify its widespread adoption.

Client Selection and Types of Assistance

Some program designs work with different clientele and some programs use screening mechanisms or criteria to select their clients. However, the programs we have observed have generally accepted all comers, in effect allowing for clients to self-select. In general, the evidence suggests that this can work quite well. Thus, Chrisman, Carsrud, DeCastro and Herron (1990) found that both males and females benefited equally from SBDC assistance. Similarly, Chrisman and Carsrud (1991) found that minority clients were also able to capture commensurate levels of benefit from that program. Further analysis by industry suggests that special program designs are not necessary for value to be added to client businesses (Chrisman 1989).

On the other hand, different types of assistance are often necessary for clients in different situations. Chrisman and Danforth (1995) have shown that the one variable that does appear to make a substantial difference is the strategy of the venture. They found that ventures utilizing cost leadership strategies gained benefits primarily from a mix of strategic and administrative assistance, whereas ventures utilizing differentiation strategies gained the most from a combination of strategic and operating assistance. Considering these collected studies, it seems to makes more sense to design programs based on the stage of development or overall strategy of the business than on gender or ethnic background of the entrepreneur or even on the industry sector of the venture.

Assistance as a Knowledge Resource

Those new to the research may be surprised at the degree of impact of entrepreneurial support programming. A limited number of hours of intervention, particularly but not exclusively in early-stage ventures (pre-ventures), appears to have an impact that is clearly discernible for at least three to five years (Chrisman and McMullan 2000). The question is, why? There are a number of likely explanations. Chrisman (1999) and Chrisman and McMullan (2000) proposed that the tacit knowledge clients gain through the "learning by doing" approach utilized by SBDC counselors provides them with a sustainable competitive advantage. Those authors suggest that this advantage may explain why SBDC clients have higher start-up rates, lower failure rates, and achieve a greater average size after a period of time than entrepreneurial ventures in the general population. McMullan and Long (1990) propose a theory of entrepreneurship with opportunity identification as the cornerstone. They argue that it is the fundamentals of the opportunity (what, where, who, and why) that have the most powerful and lasting effect on the subsequent development and ultimate success of the venture. These complementary theories essentially support the concept and importance of strategy as the matching of opportunity and resources. As the empirical studies of Chrisman (1989), Chrisman and Carsrud (1989), and Chrisman and Danforth (1995) point out, outsiders can be particularly effective when they help clients focus on essential strategic issues.

Sources of Service Provision

For many people the institutional setting for entrepreneurial support programming will be of utmost concern. Should these programs be the province of government, non-government organizations (NGOs), universities, or private business? It may be true that some program designs may be better suited to certain delivery mechanisms than to others. The fact that all of the above designs have been used in universities suggests that university-based designs can work well. However, some Small Business Development Centers are run out of government offices. The 13-week course-based design (Fastrac/WEN) has done well under an NGO model. None of these programs is operated out of the private sector, although similar programs have been run by private enterprise (but not publicly evaluated). In summary, it appears that the institutional format is less important than might be generally assumed. Perhaps, however, it is still important to maintain substantial latitude for local experimentation as is found within the SBDC programming.

At the micro level, those executing the programs need to be experienced, educated practitioners and not just academics (including teachers in Master's Degree programs). Birley, Myers, and McMullan (1993) found entrepreneurship educators to be much more enterprising than might commonly be expected. McMullan and Gillin (1998) recognized that the combination of practitioners and academics was fundamental to the success of the Swinburne program. With one exception, all educators teaching in this graduate program had noteworthy entrepreneurial experience. Both the SBDC and the Fastrac/WEN programs rely heavily on practitioners with relevant education in business and entrepreneurship. The SBDC often accepts experienced graduate students and some experienced faculty into its programming.

Conclusions

Research on the effectiveness of entrepreneurial support programming is a relatively new phenomenon. In the last 15 years, some designs have been repeatedly studied and have consistently demonstrated favorable results. This article has focused on programs that have generated supportive outcome data. The various entrepreneurial support programs we have reviewed provide different costs and benefits to different segments of the entrepreneurial community. We believe that there are opportunities to adopt and adapt these program designs in western Canada; however, there are numerous designs from other parts of the world that may be worth modifying. For example, a training program for early-retired executives has produced good early results in Great Britain (Watkins and Morris 1981; Watkins 1982).

The big-picture change is the professionalization of entrepreneurship.[9] As long as the best of new research ideas are passed on to programmers, administrators, instructors, and ultimately the clientele, then such forms of entrepreneurial support programming should continue to be useful (McMullan and Long 1990). People need to take up the challenge of promoting the professionalization of entrepreneurial support programming around the world (McMullan 1988; Vesper and McMullan 1988).

Although this article should make it clear that there are many approaches to entrepreneurial support programming that have been successful, there are some common or similar elements that can be identified. They (1) are educational in nature, (2) focus more on strategic rather than tactical issues, (3) are designed to develop both tacit and explicit knowledge about business ventures, (4) contain a strong emphasis on results-based programming where education is tied to action-based outcomes and not just knowledge acquisition, (5) are generally delivered on a just-in-time basis, (6) are universally available with emphasis on client self-selection rather than selection based on program criteria, and (7) are affiliated with universities. As further experimentation is attempted to meet the changing needs and economic situations of Canadian society, we hope that the lessons learned from successful innovations in entrepreneurial support programming will not be forgotten.

[9] Increasing knowledge in the field of entrepreneurship, along with advanced education programs such as the one at Swinburne, are paving the way toward the professionalization of entrepreneurship.

Appendix 1: Program Summaries

	Program			
	Small Business Development Center	Fastrac/Western Entrepreneurial Network	Swinburne Masters in Entrepreneurship	SBI/ Calgary Student Projects
Institutional Setting	Usually University-based	May be affiliated with university or economic development agency	University-based	University-based
Delivery Method	One-to-one counseling	13-week planning course	Master's level degree program	Projects within a course context
Service Providers	Professional counselors, professors, graduate students	Instructors with teaching and/or entrepreneurial experience	Instructors with teaching and entrepreneurial experience	Graduate and undergraduate students
Clientele	Pre-ventures; smaller growing firms	Micro-businesses, home based businesses	Students who want to start businesses	Entrepreneurs and students
Type of Counseling	Primarily strategic with some tactical	Balanced strategic and tactical	Primarily strategic with some tactical	Primarily strategic with some tactical
Knowledge Resources	Emphasis on tacit with some explicit	Emphasis on explicit with some tacit	Balance between tacit and explicit	Tacit for students; explicit for clients
Outcomes	Higher growth, survival, innovation, financing; high ratio of benefits to cost of program measured in tax revenues	Higher growth, confidence, business preparation and knowledge acquisition	Very high proportion of students start technology-based growth businesses; high initial employment levels	High growth and value added for clients; Preparation for entrepreneurial career for students with modest increase in start-up activity

References

Bartik, T. J. (1994). Better evaluation is needed for economic development programs to thrive. *Economic Development Quarterly* 8(2): 99-106.

Birch, D. (1979). *The Job Generation Process*. Cambridge: MIT Program on Neighborhood and Regional Change.

Birch, D. (1987). *Job Creation in America*. New York: The Free Press.

Birley, S., Myers, A., and McMullan, W.E. (1993). Academic entrepreneurs. In *New Directions in Small Business Research*, ed. Ratkin, E. Chall, and C. Mason. Aldershot, U.K.: Avebury.

Brown, C.A., Colborne, C.H., and McMullan, W.E. (1988). Legal issues in new venture development: A survey of entrepreneurial behaviour. *Journal of Business Venturing* 3(4): 273-86.

Chrisman, J.J. (1989). Strategic, administrative, and operating assistance: The value of outside consulting to pre-venture entrepreneurs. *Journal of Business Venturing* 4: 401-18.

Chrisman, J.J. (1997). Program evaluation and the venture development program at the University of Calgary: A research note. *Entrepreneurship Theory and Practice* 22(1): 59-73.

Chrisman, J.J. (1999). The influence of outsider-generated knowledge resources on venture creation. *Journal of Small Business Management* 37(4): 42-58.

Chrisman, J.J., and Carsrud, A.L. (1989). Outsider contributions to the development and change of business ventures. *Journal of Organizational Change Management* 2(3): 75-87.

Chrisman, J.J., and Carsrud, A.L. (1991). Outsider assistance needs of pre-venture and established small businesses: A comparison of minority and nonminority clients. *Entrepreneurship and Regional Development* 3: 207-20.

Chrisman, J.J., Carsrud, A.L., DeCastro, J., and Herron, L. (1990). A comparison of assistance needs of male and female pre-venture entrepreneurs. *Journal of Business Venturing* 5: 235-48.

Chrisman, J.J., and Danforth, G.W. (1995). Strategy as a determinant of the perceived value of outsider assistance to new ventures: An exploratory study. *Journal of Small Business Strategy* 6(2): 47-68.

Chrisman, J.J., Hoy, F. and Robinson, R.B., Jr. (1987). New venture development: The costs and benefits of public sector assistance. *Journal of Business Venturing* 2: 315-28.

Chrisman, J.J., Hoy, F., Robinson, R.B., Jr., and Nelson, R.R. (1987). Evaluating the impact of SBDC consulting: A reply to Elstrott. *Journal of Small Business Management* 25(1): 72-75.

Chrisman, J.J., and Katrishen, F. (1994). The economic impact of Small Business Development Center counseling activities in the United States: 1990-1991. *Journal of Business Venturing* 9: 271-80.

Chrisman, J.J., and Leslie, J.D. (1989). Strategic, administrative, and operating problems: The impact of outsiders on small firm performance. *Entrepreneurship Theory and Practice* 13(3): 37-51.

Chrisman, J.J., and McMullan, W.E. (1996). Economic theory, empirical evidence, and the evaluation of small business assistance programs: A reply to Wood. *Journal of Small Business Management* 34(2): 56-66.

Chrisman, J.J., and McMullan, W.E. (2000). A preliminary assessment of outsider assistance as a knowledge resource: The longer term impact of new venture counselling. *Entrepreneurship: Theory and Practice* 24(3): 37-53.

Chrisman, J.J., and McMullan, W.E. (2002). Some additional comments on the sources and measurement of benefits of small business assistance programs. *Journal of Small Business Management* 40(1): 43-50.

Chrisman, J.J., McMullan, W.E., and Vesper, K.H. (1998). *Western Entrepreneurial Network: Review of Programming*. Denver: US West.

Chrisman, J.J., Nelson, R.R., Hoy, F., and Robinson, R.B., Jr. (1985). The impact of SBDC consulting activities. *Journal of Small Business Management* 23(3): 1-11.

Elstrott, J.B. (1987). Procedure for improving the evaluation of SBDC consulting activities. *Journal of Small Business Management* 25(1): 67-71.

Jackson, W.T., Vozikis, G.S., and Babakus, E. (1992). SBI intervention: An old problem – a new perspective. *Journal of Small Business Strategy* 3(2): 15-30.

McMullan, W.E. (1988). The economics of entrepreneurship education. *Journal of Small Business and Entrepreneurship* 6(3): 8-18.

McMullan, W.E., Chrisman, J.J., and Vesper, K.H. (2001). Some problems in using subjective measures of effectiveness to evaluate entrepreneurial assistance programs. *Entrepreneurship Theory and Practice* 26 (1): 37-54.

McMullan, W.E., and Gillin, L.M. (1998). Developing technological start-up entrepreneurs: A case study of a graduate entrepreneurship education program at Swinburne University. *Technovation* 18(4): 275-87.

McMullan, W.E., and Gough, V. (1999). Developing entrepreneurial MBAs: A case study in Calgary. *Proceedings of the Canadian Council of Small Business and Entrepreneurship.*

McMullan, W.E., and Long, W. (1987). Entrepreneurship education in the nineties. *Journal of Business Venturing* 2(3): 261-75.

McMullan, W.E. and Long, W. (1990). *Developing Entrepreneurial Ventures*. San Diego: Harcourt Brace Jovanovich.

McMullan, W.E., Long, W.A., and Graham, J.B. (1986). Assessing economic value added by university based new venture outreach programs. *Journal of Business Venturing* 1(2): 225-40.

McMullan, W.E., and Vesper, K.H. (1987). New ventures and small business innovation for economic growth. *R&D Management* 17(1): 3-14.

Monroe, S.R., Allen, K.R., and Price, C. (1995). The impact of entrepreneurial training programs on transitioning workers: The public policy implications. *Frontiers of Entrepreneurship Research.* Babson Park: Center for Entrepreneurial Studies, Babson College.

Monroe, S.R., Price, C., and Burns, M. (undated). *Homebased businesses: The incubators of global growth*, unpublished paper.

Monroe, S.R., Price, C., and Prinster, C. (1996). The triad theory essential to entrepreneurial success. *Frontiers of Entrepreneurship Research*, Babson Park: Center for Entrepreneurial Studies, Babson College.

Price, C., Allen, K.A., and Monroe, S.R. (undated). *The effect of geographical and socio-economic differences on the replicability of entrepreneurial training programs*, unpublished paper.

Price, C., Burns, M. and Monroe, S. (undated). *Growing home-based businesses with entrepreneurial training: The Premier FastTrac Program*, unpublished paper.

Price, C., and Monroe, S. (1993). Educational training for women and minority entrepreneurs positively impacts venture growth and development. *Frontiers of Entrepreneurship Research*, Babson Park: Center for Entrepreneurial Studies, Babson College.

Price, C., and Fleming, D. (undated). *Four year study of Colorado entrepreneurship with minority and women business owners*, unpublished.

Price, C., Monroe, S., and Larsen, G. (1992). The positive effects of economic development by training women and minority entrepreneurs. *Frontiers of Entrepreneurship Research*, Babson Park: Center for Entrepreneurial Studies, Babson College.

Solomon, G.T., and Weaver, K.M. (1983). Small Business Institute economic impact evaluation. *American Journal of Small Business* 8(1): 41-51.

Vesper, K.H., and McMullan, W.E. (1988). Entrepreneurship: Today courses, tomorrow degrees? *Entrepreneurship: Theory and Practice* 13(1): 7-13.

Walker, M. (1999, November 19). Flaws mar examination of Alberta oil royalties. *Calgary Herald*: A25.

Watkins, D. (1982). Use of business competition to stimulate business formations. *Frontiers of Entrepreneurship Research*, Wellesley: Center for Entrepreneurial Research, Babson College: 469-84.

Watkins, D., and Morris, J.F. (1981). U.K. government support for entrepreneurship. *Frontiers of Entrepreneurship Research*. Wellesley, MA: Center for Entrepreneurial Research, Babson College.

Watts, L.R., and Jackson, W.T. (1995). The SBI program and student outcomes: A study of business policy classes. *Journal of Small Business Strategy* 6(1): 93-103.

Weinstein, A., Nicholls, J.A.F., and Seaton, B. (1992). An evaluation of SBI marketing consulting: The entrepreneur's perspective. *Journal of Small Business Management* 30(4): 62-71.

Wood, W.C. (1994). Primary benefits, secondary benefits, and the evaluation of small business assistance programs. *Journal of Small Business Management* 32(3): 65-75.

Wood, W.C. (1999). Benefit measurement for small business assistance: A further note on research and data collection. *Journal of Small Business Management* 37(1): 75-78.

Chapter 10

Developing Entrepreneurs in a Hybrid Management and Entrepreneurship MBA: A Case Study in Calgary

W. Ed McMullan[1] *and Vance Gough*

Abstract

The University of Calgary has been teaching MBA courses in entrepreneurship since 1976. In 1993, the traditional, full-time MBA was changed to an MBA in Enterprise Development. This University of Calgary program was conceived as a compromise or hybrid program designed to develop both managers (preferably innovative) and entrepreneurs (preferably growth). The new program provided an opportunity to explore whether entrepreneurs and managers could be effectively educated together using a modified MBA format.

The jobs of graduates before and after the introduction of an Enterprise Development MBA were traced. In one year, the rate of new business formation and development went from 14%[2] *(a level equivalent to that of the Canadian population at large) to 38%.*[3] *This intriguing change in results came despite very manageable changes in the overall operation of the program with*

1 Thanks to: Karl Vesper for his provocative questioning and for selective editing; to Debi Andrus, Nicole Coviello, Rob Isaac, Ron Murch and Amy Pablo for their editing assistance; to Dana Wilson for her persistence in getting a 96% response rate and to Stacey Strike for her assistance with data presentation.

2 If this data is adjusted for some of the non-comparable aspects then a more realistic measure of change in post-graduate, new business formation and early stage development between the two programs goes from 5.65% under the traditional MBA to 38% under the Enterprise Development MBA — a 673% change.

3 This program produced significant changes from the University of Calgary's earlier programs (673% increase in entrepreneurial activity). However, comparative to other programs, such as Swinburne, and with regard to consistent results over time, the results are modest.

similar students, the same faculty, and the same entrepreneurial support system.

For the main period studied, 67% of the graduates were in the roles that might broadly be classed as either intrapreneurial or entrepreneurial. The remaining 33% were divided: 27% in non-entrepreneurial corporate roles and 6% other.

Insiders recognized that the program was subject to drifting back toward MBA norms. By implication, the hybrid form of entrepreneurship education, at least as it has been practised at the University of Calgary, has produced both modest and inconsistent results over time and therefore may not be a robust form of in-depth entrepreneurial education.

Introduction

In 1993, the University of Calgary decided to modify its MBA program so as to develop more entrepreneurial talent in its graduates. The desired outcome of the program was "entrepreneurial leaders." The mission statement was to develop graduates who "create and grow their own enterprise or lead and sustain growth and development in an existing enterprise."

There were three defining features of the new Calgary Enterprises Development MBA, all three of which were uncommon in MBA programs in North America. The first was the entrepreneurial orientation of the program as a whole: many schools had concentrations or majors in entrepreneurship, but few had entrepreneurship integrated throughout the program as a whole. The second feature was a heavy concentration of an applied experiential teaching methodology based on the use of live cases. The third key feature was an integrated, team-teaching approach that was employed throughout the first year of the program.

The entrepreneurial approach was implemented at Calgary in five ways: First, changing the MBA program's stated goals; Second, introducing some limited entrepreneurial material into the program core; Third, shifting more of teaching to live cases; Fourth, selecting functional area content of most use to entrepreneurs; Fifth, organizing the functional area materials by stage of business development. The subject matter of the program was also frequently taught in modules rather than in standard length courses, so as to accommodate the different learning requirements of entrepreneurial graduates.

The applied project approach to education was introduced mainly through the MBA clinics. The clinic program was comprised of two full-year courses over the program's two-year duration. In the first year the students, in teams of four or five, worked with mid-sized entrepreneurial companies from the Calgary business community.[4] The firms typically had from 25 to 150 employees and were big enough to have the functional area differentiation needed to understand course materials, but small enough that students could grasp the entirety of the client's strategy. The client firms became the medium for appreciation and understanding of the concepts and theories taught in the functional area courses. Students commented in their end-of-year reviews and feedback sessions on the MBA program that they spent as much as 60% of their time and effort during the first year analyzing, studying, and applying academic concepts to the client firms. In the second year, the students not only typically consulted in teams with four-early stage ventures, but they were also encouraged to pursue an entrepreneurial initiative of their own. The project options included: (1) to launch a start-up, (2) to help the development of a smaller, growing firm, or (3) to contribute creatively and intrapreneurially to the development of a mature corporation.

An integrated teaching approach was used in the first year of the Enterprise Development MBA. All faculty members worked together as a team both to plan and execute the program. Although the program content required no formal courses in entrepreneurship, the functional area courses were partially organized by stage of business development, from start-up to growth to the development of larger firms. In addition, several option courses were available in the second year to those wanting a more in-depth treatment of various aspects of new venture development.

Few of the program's first-year integrated teaching faculty had experience as entrepreneurs. The two professors with entrepreneurial experience were responsible for one-fifth of the programming during the main period under study. Other faculty members were building experience bases with exposure to entrepreneurial project clients and through independent consulting. Integration ended in the second year of the program, when students selected their own courses and timetables and were required to take one course, the Clinic session, as part of the Enterprise Development MBA.

---

[4] Each client firm paid $5000 to have students provide an audit of their business over the fall and with the business's expansion strategy, typically over the winter.

The teaching method of the clinical and integrated portions of the program was learning-by-doing (i.e., experiential learning). It was live and project-based, providing learning that was more easily transferable to real business situations. In addition, students were exposed to substantially more consulting activity than they might otherwise have been.

The Enterprise Development MBA was, first of all, applied and hands-on. Students consulted[5] for their first year with one mid-sized business and completed projects in all of their functional area courses related to that company. During the first year, the program was integrated and team-taught at substantial extra cost to the Faculty of Management.[6] The second year was more typical of MBA programs elsewhere, with the exception that students took a full-year clinic typically involving consulting with 4 to 6 newer, smaller ventures. Second, the program was entrepreneurial, in that entrepreneurs were brought into classes as speakers and both cases and projects often had an entrepreneurial bent in the first-year courses. In the second year, students were typically encouraged to develop their own ideas for businesses.

Entrepreneurial Graduates from a Conventional MBA Program

To establish a baseline of entrepreneurial activity, Daniel Dang and Ed McMullan (1994) began a study of MBA graduates for the period during which the MBA program offered only second-year option courses in entrepreneurship as the prime strategy for encouraging entrepreneurial behaviour. Of the nearly 1000 graduates during that period, 150 of the graduates known to be engaged in entrepreneurial activity were contacted by telephone or fax or both. The sample was collected in two stages. From alumni records and other sources, data was available to identify those who were self-employed, running their own business, or holding a sizeable equity stake (>5%) in a business in which they held an executive position. In the first stage, information was persistently solicited from about 100 pre-indicated, graduate entrepreneurs to get as complete a census of this sub-population as possible. In the second stage, a sample of 50 was drawn randomly from the remaining graduate

[5] Consulting is probably the easiest way to describe what students are doing. In part, however, the consulting model was fashioned after the clinical training model used in medical faculties. The clinic model employed is probably a hybrid of business consulting activity and a medical style clinic.

[6] Faculty typically got twice the course credit for teaching in the program because of the extra time required for group meetings.

population. The percentage of those remaining graduates engaged in entrepreneurial pursuits was used as an estimate of the entrepreneurial participation rate of the rest of the population. It was found that, between 1974 and 1993, only 14% of the graduates of Calgary's MBA program either started their own companies or took senior executive positions in smaller businesses. Some selected findings are as follows:

Entrepreneurial participation rates

- The proportion of Calgary MBA graduates who were employer entrepreneurs (7%) was the same as the proportion of employer entrepreneurs in the workforce in Canada (7%).
- The number of Calgary MBA graduates who were self-employed without employees (7%) was approximately the proportion of self-employed in the workforce without employees (6.3%). (No attempt was made to determine what proportion of those contending to be self-employed without employees were better classified as unemployed passing themselves off as consultants. Many volunteered a preference to have a job.)

Investments and earnings

- The average employer entrepreneur invested $33,000 of his or her own money in the business. The average 1994 value of these entrepreneurs' investments in their businesses was between $359,000 and $401,000.[7] Average annual earnings of Calgary MBA employer entrepreneurs were $81,300 per year which was above the $71,000 average annual earnings for Calgary MBA graduates in 1994 (from alumni data).
- Both earnings and equity valuation were positively associated with the size of the business (number of employees), the number of prior business ventures and the choice of industry. Conversely, earnings and equity evaluation, alternatively, were inversely related to the percentage of the company's share equity held.

---

[7] Two methods of valuation were used. The first was four times company earnings divided by the percentage of business owned. The second was the graduates' subjective evaluation of the value of their holdings.

- Unrelated to earnings and equity were uniqueness of the products and years of work experience prior to start-up.

Business demographics

- Average age of employer entrepreneur businesses was 8.6 years.
- Industry breakdown of businesses was as follows: professional consulting (49%), oil and gas development (15%), manufacturing (9%), wholesaling, including oil and gas marketing (7%), retailing (4%), publishing (2%), restaurants (2%) and miscellaneous (12%).

Entrepreneurial strategies

- Route of entry into employer entrepreneurship: start-ups (65%), bought-in (12%), family succession (10%), turnarounds (7%) and spin-offs (5%).
- Average hours worked per week: 52 hours.

The self-employment rate of Calgary MBAs (under the old and more typical MBA program) was no greater than that for the Canadian population at large, even though the University of Calgary had an active concentration in entrepreneurship.[8] The entrepreneurship concentration appears to have had little impact on the rate of graduate entrepreneurial activity. Part of the explanation for this relatively low rate of entrepreneurial involvement among the graduates might be that, until 1986, the MBA was almost exclusively a part-time program filled with students who were full-time employees, many of whom were working in the high-wage local oil and gas sector. Another part of the explanation may be that the concentration in entrepreneurship was more a collection of courses open to those interested or curious about the topic than a required set of courses needed for a minor or major in the subject. Thus, despite the fact that many of the students came from outside Calgary and appeared to have been initially attracted by entrepreneurship programming, by the time they entered their second year, few took

[8] The courses in entrepreneurship began in1976 at Calgary and a four-course concentration was underway by 1980. Thus the graduates from 1974 through 1979 were unable to obtain a full concentration in entrepreneurship.

more than one or two courses in entrepreneurship.[9] In all likelihood, those who had a strong interest took more courses. Possibly, the courses served the interests of the 7% to 14% of the population who were already predisposed to entrepreneurship.

Defining Features of Calgary Enterprise Development MBA Graduates

Survey of the Classes of 1995 and 1996

A telephone survey was conducted with both 1995 and 1996 graduating classes, respectively two years and one year after graduation. Repeated calling allowed the researchers to obtain a 96% sample of 68 graduates (33 graduates in 1995 and 35 in 1996). All contacted were cooperative and helpful in providing the data requested. The graduates were asked about their career paths since graduation, future career objectives, consulting activities, job responsibilities, and about the ventures they had created. The process of classification was complex. After the surveys were completed, the authors found that their classification system was ineffective, and as a result they were forced to create new categories to more closely match the activities of the graduates. This creation of new categories involved much discussion of both class definitions and of respondent fit. Where there was ambiguity, the usual choice was to err by classifying people in less entrepreneurial categories.

Roles in business, now and later

At the time of the survey, 67% of respondents were in entrepreneurial roles, entrepreneurial support roles or full-time consulting roles. A further 27% were in non-entrepreneurial roles. The remaining 6% were in school, parenting, or unemployed.

In addition to questions about their current jobs, graduates were also asked about their career aspirations. Sixty-four percent of graduates expressed an intention to be in an entrepreneurial or consulting role within five years. For example, when those who wanted to be in their own businesses (other than consulting) in the future were added

[9] The fact that the first year of the program had the tendency to channel students in directions other than entrepreneurship may be in part due to the lack of entrepreneurial orientation in the first year prior to the new program.

to those already in their own businesses the percentage increased from the current 14% to an anticipated 28%. Of those in independent consulting businesses (14%), only half wanted to be similarly employed within five years. The number of graduates wanting to become members of an executive team in a smaller, growing firm would increase from the current 6% to an anticipated 12% within five years, while those wanting to be in entrepreneurial or partially entrepreneurial roles in large corporations drops from the current 21% to an anticipated 10%. Most respondents who were currently in non-entrepreneurial roles wanted to stay in these types of positions.[10] Two percent of respondents were currently in other roles (such as attending school or full-time parenting), but this figure was expected to increase to 10% within five years. A possible interpretation of these changing patterns of career objectives may be related to the issue of financial risk. Some may wish to engage more risky entrepreneurial options by degree. By starting with consulting or with an entrepreneurial position in a larger firm, graduates may be preparing themselves over time to take a more controlled, higher-stakes gamble on independent non-consulting ventures later.

Role classification

- *New venture creators* in non-consulting businesses (14%) included people who have started different types of businesses. The businesses included the following: a natural bread store, a movie theatre, a women's entrepreneurship program, five bagel shops with franchising, a health care program for chiropractic services, and a native arts retail store.
- *Founders of consulting firms* (14%) included those people who have started new consulting businesses. This group was handled separately, because there is a tendency for unemployed professionals, MBAs included, to call themselves consultants. Special care was made to insure that each graduate in this category was not only purposefully choosing to be a consultant but was also demonstrating success in getting contracts.
- *Executives in small growing firms* (10%) included those indi-

[10] There was only a 1% difference between current and anticipated involvement within a five-year time span.

viduals who have joined a team of people starting newer, small businesses, but not at their outset. They typically have equity or the promise of equity as a significant incentive to perform. Members of this category were involved in the following industries: wine distribution, drugstores, organic soap manufacturing, industrial design service, and golfwear manufacturing and distribution.

- *Corporate entrepreneurs* (9%) included people involved in a number of different types of businesses, including the following: telecommunications, biotechnology commercialization, and trucking and trucking logistics.
- *Entrepreneurial support positions* (4%) included a number of people who have taken jobs wherein their primary responsibility is to work with entrepreneurs, assisting them in some way or another. Graduates in this category included a commercial lending manager in charge of small business lending, an agricultural product development specialist, and a lender of last resort for small businesses.
- *Consultants in consulting firms or corporations* (4%) mostly included graduates who acquired higher-prestige positions in the local offices of large consulting firms.
- *Partially entrepreneurial corporate roles* (12%) included graduates who saw themselves in roles that were largely or completely entrepreneurial but did not involve the formation of new business entities. Classification was further assisted by selected statements, such as: "developing new systems," "putting in new programs," "business development responsibilities," "sets corporate direction and develops new business," "in charge of customer development," "acquisition of new companies," "manages a number of small clinics," "set up new look in gas bars," "new product issues are central," and "clients are her own, making it like her own business."
- *Non-entrepreneurial corporate roles* (27%) included some graduates who were not only in non-entrepreneurial roles at the time of the study but also never wanted or intended to be in entrepreneurial roles. Some saw the roles as partially entrepreneurial but did not provide sufficient evidence for the researchers to share their self-classification. A few

were forced into such positions out of financial necessity and saw themselves becoming entrepreneurial when future circumstances permitted. One of the most interesting findings about this group was that more of these people are in line positions (55%) as opposed to staff positions (45%). People in this category included: financial analysts (3), account managers (3), health planning specialist, research analyst, financial services manager, and one person in sales and promotion.

- *Other* (6%) includes graduates still pursuing their education (3%), consulting on a part-time basis (1%) or currently unemployed (4%).

Consulting after graduation

Seventy percent of graduates indicated that they had done consulting since they graduated. An incomplete sample of reported consulting projects indicate the variety of consulting specialities:

Venture performance to date

Of those graduates who had started new non-consulting ventures, 62% claimed that their business was in excellent financial condition, while 19% claimed theirs was in good condition. Only 13% reported their business to be in poor financial health. Furthermore, 24% claimed to be making more than they would earn elsewhere, 29% felt they were making about the same, 6% felt they were earning less than they would earn in another job, and 41% claimed to be earning much less. In summary, the subjective reports of financial performance seem surprisingly good, given the time it takes for most businesses to make money and the fact that few of the reporting businesses are more than two years old.

Innovations in industry and community

With regard to innovations, at least within their industry and community, a high percentage of graduates who started businesses claim to have introduced a wide variety of innovations:

The intensity of reported innovations across the spectrum of different disciplines suggests that students are putting into effect what they are learning from their various functional area courses.

Table 1

Type and Number of Consulting Projects Completed
Since Graduation

Type on project	# of projects
General business consulting	5
Marketing research	4
Marketing and market planning	4
Start-up consulting	4
Business planning	3
Finance and financial analysis	3
Accounting and accounting software	2
Business development	1
Human resources	1
Fundraising	1

Table 2

Innovation Introduced by Graduates
Starting Non-Consulting Businesses (14% of sample)

Innovation introduced	% of respondents
New or improved product	70%
New markets	60%
New channels of distribution	50%
New marketing or sales methods	60%
New or improved services	50%
New production methods	45%
New methods of management	55%
New methods of organization	45%
New production methods	45%
New methods of finance	25%

Vignettes on the Program's Graduates

The following five vignettes of graduates have been selected to illustrate the program participants and how they were influenced by the Enterprise Development MBA Program.

John Pallas (Bachelor of Physical Education) had been in sports administration for three years, and was in an entrepreneurial role making money through events management and promotion. He wasn't pleased to be working hard and making lots of money for a not-for-profit organization while pulling down a small salary for himself. At 25, he saw the MBA as a tool for taking control of his own destiny and entered the program with some business ideas in mind. The program was so overwhelming with so much good material that his business ideas were put on the back burner. The opportunity to work with different clients fuelled his entrepreneurial aspirations even more. "Being thrown into the fire at the outset was a good idea." He hadn't thought of starting his own consulting firm, but the idea flowed naturally out of the in-program consulting experience: "It just made sense." Early on, the idea of working for a big consulting firm began to draw him, but the further he got into the program the more he changed his mind about working for someone else: "I pulled a 180 on that." By the second year of the program, he was making nearly as much from consulting as he had made at his last job, and he was still going to school and getting course credits, getting work experience, and networking. "The relationships that I formed during my consulting at school just continued afterward. I make three times what I made before the MBA using 3/4 of my time and my time is very flexible." He further acknowledged that he was making somewhere between 25% and 50% more than his peers who took more conventional jobs in industry. "Moreover, much of my time is working in an industry that I would never have considered before the MBA." At the time of the survey he claimed that he had a "fire burning about doing his own thing" and was open to a wide variety of opportunities. John's role was classified as a *"founder of consulting firm."*

Dave Bell (Bachelor of Arts in Economics) had started three small, marginal businesses in Toronto prior to entering the MBA program: a part-time washroom advertising business, a footware distribution business, and an adventure canoe trip business. He decided on attending the University of Calgary because of its entrepreneurial orientation and its specialization in tourism. He thought that he might build his adventure canoeing business into something bigger. Overall he was hoping that the MBA would give him more knowledge on how to finance and grow a business.

"First year was outstanding in the exposure to business overall. I found it of noteworthy value to study cases and discuss them in classes with my fellow students and professors. Second year left me more to my own initiative. The program allowed me to do due diligence on my canoeing idea which I ended up dropping...Before the program I used to think that entrepreneurship meant finding a neat product idea and running with it. The program gave me a much more integrative appreciation of all the different things one has to do to succeed in business. For my experience combined with my goals it was a perfect program."

"One of my classmates introduced me to my current boss – a husband and wife team. They wanted to retire and were looking for a general manager to take over." As it turned out, in first year, Dave's client company was in the garment industry. At the time of the study he was managing a garment company with 26 employees, $2.9 million in sales, and a current growth rate of 45%. "I participate in profit-sharing and there is an opportunity for me to buy an ownership position out of my profit share." Dave's company was purchased by a larger firm from Toronto. He saw even more opportunities in the firm for himself as a result. "For me it is a dream job. It's great." Dave's role was classified as *"executive in small growing firm."*

Christine Nehring (Bachelor of Music) was working in an administrative capacity for two years prior to entering the MBA program at 26. She had been struggling over a four-year period with the idea of giving up music as a career and obtaining some other form of professional credential. She liked the applied nature of the Calgary program although she "was a little nervous about its entrepreneurial orientation." The MBA made her more knowledgeable as to how to become an entrepreneur if that was what she wanted. "The program really demystified entrepreneurship for me. Getting to know the clients made me realize they were like me. Watching my fellow classmates who were starting businesses further helped me understand the process." However, for her the biggest impact of the program was in validating her capability to perform effectively in a business environment. It gave her the tool set she needed. At the point of finishing the program she still had a relatively limited business experience base and she didn't have an idea for a viable venture. Perhaps most importantly, her husband had been out of work for some time and they needed a secure income. As a result she looked for a job in a large business. At the time of the study she was working "in a marketing capacity in the fertilizer business" in Calgary for one of Canada's biggest companies. She worked on anything from establishing rates to helping her customers penetrate new markets. From Christine's perspective the job was not particularly entrepreneurial and that was not a problem for her. She liked her situation "just fine" and saw herself similarly employed for the foreseeable future. Christine was classified as *"employee in non-entrepreneurial corporate role."*

Kim Molloy (Bachelor of Arts in Psychology) worked for two years at Proctor and Gamble as a sales representative before starting the MBA at 25. She realized that in order to be on the fast track at P&G or any other large firm she would need an MBA. She chose Calgary because of the "better balance there between the softer subjects and the quantitative disciplines." Initially she thought she would specialize in Human Resources. Largely because of the work with clients in both the first- and second-year clinics, she became increasingly attracted to setting out on her own. "I got more of a grasp of what an entrepreneur was when they profiled entrepreneurs in class and I realized that I matched that profile." She found the applied learning permitted by working with clients while taking courses valuable. Also beneficial was the productive small group sharing of ideas with classmates. At survey time she worked as an account executive in a mid-sized firm of about 400 employees. "The president is very entrepreneurial. The company is entrepreneurial. You cannot work here unless you are entrepreneurial." Her work requires an understanding of a number of disciplines: marketing, finance, computer systems and contract law. "It is as if you are responsible for a certain piece of business and it is considered yours. There is a sense of personal ownership. In the future if I were to go elsewhere it would be for much the same thing as I have here or to do my own business. I do not think that I could work for a more conventional large business." Kim's role was classified as *"employee in partially entrepreneurial corporate role."*

Carolyn McConnell (Bachelor of Human Ecology) was a nutritionist for the Inuvik Regional Health Board. Her increasing frustration that her specialty area of preventive medicine was continually being cut led her to look at other options. She felt that an MBA might be used in conjunction with her undergraduate knowledge and experience to help her attain an administrative level in the health industry.

"In retrospect, the biggest value in the program to me was in working with entrepreneurs and seeing how they think, although I didn't begin to realize that until towards the end of the program. My first clue came from employers who were impressed by my having experience working with entrepreneurs. I think the project that I did with two other students looking into starting a cooking school in second-year clinic was useful in making me more entrepreneurial."

She also advocated that students get involved in extracurricular activities because she had found them particularly valuable. These activities not only helped develop her networking skills but also helped her learn such things as making cold calls and getting people to volunteer.

Because of all the connections with the business community she felt that "students who intend to stay in Calgary are at an advantage for realizing the full potential of this type of an MBA." At survey time Carolyn was a commercial lending officer with the Business Development Bank of Canada in Yellowknife. "My consulting experience during the MBA was the main factor influencing their interest in my application, as was the MBA itself." Carolyn's role was classified as *"an employee in an entrepreneurial support position."*

Survey Results from the 1998 Graduating Class

By 1997, the faculty teaching in the Enterprise Development MBA recognized that the program had slipped somewhat back toward a traditional MBA. The first-year courses were no longer taught in three phases from start-up to growth to large enterprise, and the entrepreneurial content within the courses appeared to be diminished. This drifting back toward the standard MBA model of educating middle managers had to be fought year after year each time new faculty were recruited into the program. A follow-up study of the 1998 graduating class in 1999 demonstrated that the number of students with jobs had increased to 90% and those who were self–employed was down to 5%.

Discussion

A few general conclusions can be drawn from the study data. First, a concentration of entrepreneurship courses offered at Calgary over the better part of an 18-year period did not appear to increase the rate of entrepreneurial activity among graduates.[11] Second, in the first two graduation classes, the new Enterprise Development MBA did appear

[11] The proportion of only about 14% of traditional MBAs developing their own business may be viewed as low when compared to the Canadian population at large, given that MBAs are trained in business and entrepreneurship to a much greater degree than the Canadian population at large. If the comparison point were percentage of university graduates who go into their own business, then one might have drawn different conclusions:
"As the then Minister for Small Firms in the Department of Employment (for Great Britain) observed in 1987, only one-third of 1% of British graduates go into self-employment, compared with 2% in the United States and 2.5% in Japan." (Kirby, 1992). If 1% to 2% were the base expectation for Canadian graduates entering self-employment, then a concentration producing 14% self-employed would have appeared to have increased the participation level.

to increase the amount of entrepreneurial activity one and two years after graduation.[12] Third, those who wanted more traditional jobs were also generally able to get the type of jobs they wanted from the new program. Fourth, the program drifted of its own accord back toward a traditional MBA format and the entrepreneurial results appeared to decline accordingly. It may also be noteworthy that the year in which the proportion of entrepreneurial graduates first increased did not appear to mark any related change in the degree which the Calgary climate encouraged or supported entrepreneurial activity.

[12] One's interpretation of the size of the change from before to after the introduction of the new program might be different if one were to integrate three additional pieces of information into the analysis of the change. An argument could be made to adjust the statistics from the 18-year period prior to the new program for three considerations, namely: (1) the average period from graduation to the creation of the reported enterprise was 3 years (whereas the new program study looks at graduates 1.5 years beyond graduation), (2) about 45% of traditional MBA graduates with enterprises with employees were created prior to the program, and (3) a sizeable proportion of those claiming to be entrepreneurs without employees were more realistically classified as unemployed. To attempt to bring the pre-program statistics in line with post-program statistics, the entrepreneurs with employees are reduced by a factor of 7 to adjust for the first two observations and the entrepreneurs without employees are cut by a third to accommodate for the unemployed passing themselves off as consultants. Assuming that half of the e.e. new enterprises were created in 3 years and the other half were in six years then the 7% of e.e. ventures created by traditional MBAs would need to be divided by 4 (1.75%) to be on a comparable basis with the new sample of ventures created by Enterprise Development MBAs. Furthermore, only 55% of such ventures were created after graduation by traditional MBAs, requiring a further reduction from 1.75% to .96% to create comparability. Finally we will assume a one-third overstatement in non-employer entrepreneurs as a result of unemployed graduates calling themselves consultants in the study of traditional MBAs (from 7% to 4.69%). Therefore the comparable number of graduates among traditional MBAs producing businesses is 5.65% (i.e., 96% + 4.69%). The increase in entrepreneurial activity in a comparable 1.5 -year period from 5.65% under the traditional MBA to 38% under the Enterprise Development MBA is 673%, a truly noteworthy change.

References

Boberg, A.A., and McMullan, C.A. (1991). The relative effectiveness of projects in teaching entrepreneurship. *Journal of Small Business and Entrepreneurship* 9:14-24.

Bygrave, W.D. (1994). Doctoral students: How much have they influenced entrepreneurship education and practice? In *Frontiers of Entrepreneurship Research*, ed. W.D. Bygrave, S. Birley, N.C. Churchill, E. Gatewood, F. Hoy, R.H. Keeley, and W.E. Wetzel, Jr., 616-19. Babson Park: Babson College.

Chrisman, J.J. (1996). *Program Evaluation: A progress report on the Venture Development Program at the University of Calgary*. Calgary: Office of the Vice President (Research), University of Calgary.

Dang, D., and McMullan, W.E. (1994). Calgary MBA graduates, the entrepreneurial option and will it be rewarding. Unpublished presentation, University of Calgary.

Hornaday, J.A., and Vesper, K.H. (1982). Entrepreneurship education and job satisfaction. In *Frontiers of Entrepreneurship Research*, ed. K.H. Vesper ,526-39. Wellesley: Babson College.

Kirby, D.A. (1992). Developing graduate entrepreneurs: The U.K. Graduate Enterprise Program. *Entrepreneurship, Innovation and Change* 1(2): 165-175.

Knowles, M. (1973). *The Adult Learner: A Neglected Species*. Houston: Gulf Publishing.

McMullan, W.E., and Gillin, L.M. (1997). Entrepreneurship education in the nineties: Revisited. Paper presented at *IntEnt97*, Monterey, California.

McMullan, W.E. (1988). MBA demand for entrepreneurial education. Unpublished raw data, University of Calgary.

McMullan, W.E., and Long, W.A. (1987). Entrepreneurship education in the nineties. *Journal of Business Venturing* 2: 261-76.

Nikiforuk, A. (1997). Enterprise U: Still think that entrepreneurs are born, not made? Graduates from a unique University of Calgary MBA are proving just the opposite. *Canadian Business* 70(8): 30-34..

Porter, L.W., and McKibbin, L.E. (1988). *Management Education and Development*. New York: McGraw-Hill.

Taras, D. (1994). The impact of industrial relations strategies on selected human resource practices in a partially unionized industry: The Canadian petroleum sector.Ph.D. Dissertation, University of Calgary.

White, J. (1984). The rise of female capitalism – women as entrepreneurs. *Business Quarterly* 49(1): 133-135.

Chapter 11

A Preliminary Assessment of the Effectiveness of Project-Based Assistance to Medical Entrepreneurs at the University of Calgary

Myron L. Weber, James J. Chrisman, James O'Grady, and Vance Gough

Abstract

The Student Projects Program of the Entrepreneurship and Innovation Area of the Faculty of Management of the University of Calgary has been successful in providing assistance to organizations in fields as diverse as communications, oil field engineering, and software development. The program has been extended to medically related enterprises, suggesting that such an approach can also facilitate medically oriented entrepreneurship. The effectiveness and impact of this program on medical entrepreneurs is considered in this paper.

Introduction

In 1994, the Vice President (Research) at the University of Calgary commissioned a study of the economic benefits of faculty research (Chrisman, 1994). That study concluded that there was considerable entrepreneurial activity at the university and that there was potential for even greater numbers of business starts and other commercialization activities emanating from the research-based discoveries and inventions of faculty scientists. This was particularly true in medically oriented areas. At the same time Chrisman (1994) found that for entrepreneurship to flourish, institutional barriers needed to come down, the government needed to provide greater support and encouragement, and would-be entrepreneurial scientists needed assistance in setting up businesses.

Unfortunately, good faculty scientists are not necessarily good entrepreneurs. Technical proficiency and good product ideas must be combined with the expertise and experience necessary to create and manage an organization that can successfully develop these products and bring them to market.

Research devoted to innovation, development, and management of medical technology has so far yielded few solutions for moving from a great idea to a market-competitive, technology-based medical products company. To facilitate the success of enterprises involved in commercializing medical technology, an approach with the potential to enhance the managerial capabilities of medical entrepreneurs and/or provide ready access to such expertise is needed.

To this end, the Student Projects Program of the Entrepreneurship and Innovation Area of the Faculty of Management of the University of Calgary has, for a number of years, been successful in providing entrepreneurial assistance to organizations in fields as diverse as communications, oil field engineering, and software development. In an extensive review of entrepreneurship education, Gorman, Hanlon, and King (1997) cite the founders of the University of Calgary program as major contributors to post-secondary entrepreneurship education on both theoretical and empirical grounds. The Projects Program differs from most other programs in its cross-functional focus. Students interact with their client organizations in a wide range of functional areas as are appropriate to their clients' needs. Expertise required by student teams to deal with client problems is provided by their faculty mentors on a just-in-time basis. Although it was designed primarily to educate students about the entrepreneurial process, Chrisman (1997) found that the projects program can also add considerable value to the start-up organizations themselves.

The question is whether such a program can be successfully adapted to meet the needs of medical entrepreneurs. It is difficult to determine what characteristics of entrepreneurship are invariant and which are context-dependent (Huefner and Hunt 1994). Drucker (1985), however, contends that entrepreneurship in education or health care displays essentially the same characteristics as it does in the private sector. Huefner and Hunt reach the same conclusion in their comparison of business and consumer entrepreneurs.

The central thesis of this paper is that the projects program can provide meaningful assistance to medical entrepreneurs. Although investigation is still in an early stage, the initial evidence suggests that the lessons and successes previously experienced with other organizations have translated effectively from a number of other business enterprises to the specific case of medical entrepreneurship.

Characteristics of Medical Entrepreneurs

Researchers have identified significant contrasts between the backgrounds and values of academics and business managers. Quintas, Wield, and Massey (1992), comparing differences in process invoked, and outcomes sought, between university research and business activities, found that whereas academic knowledge is general, theoretical and long-term, business knowledge tends to be specific, applied and immediate. In particular, academic knowledge seeks to explain, whereas business knowledge is quite often tacit and focuses on the operational.

Corsten (1987), in a study of 225 small to medium-sized entrepreneurs, found significant obstacles to collaborative efforts between industry and academic researchers. He cited as the cause non-cooperative attitudes of professors who are perfectionists, lack practicality and realism, and exhibit hostility to compromise. He also reported that they ignore deadlines, reject profitability as a measure of success, are poor communicators, and ignore confidentiality requirements. He goes on to suggest that these difficulties arise from the search for scientific proof. Needless to say, these behaviors are not appreciated in business.

How the problem of culture clash between researchers and the business community is dealt with can be an important determinant of the ultimate success of many technology commercialization ventures. Foxall and Johnston (1987) pointed out that the initiators of technology invent, use, make, and apply results to their own purposes in the earliest stages. They have unique access to information on the product and intimate knowledge about the operational performance of the device. Their conclusion – that initiators of technology must be involved in the process – suggests that programs must be devised to expose researchers to, and inform them of, the role they and other actors play in the commercialization process.

Concerning the requirements for close involvement of researchers in the commercialization process, Samsom and Gurdon (1993) reported on an investigation of university scientists as entrepreneurs. Their findings suggested that an important determinant of success of faculty entrepreneurs was the acceptance of the idea that scientists can be entrepreneurs while retaining their university contacts. Such acceptance requires a substantial shift in the perceptions of both university scientists and the business community. Samson and Gurdon further observe that such acceptance might yield substantial contributions to the innovation cycles at work and in society.

In summary, challenges to effective interaction between medical and academic entrepreneurs and managers include resolving differences in education, business background, and commercial orientation. There is also a need to address the propensity of medical entrepreneurs to desire predictability in an innately unpredictable entrepreneurial situation, to prefer perfectionism over pragmatism, and to demonstrate antipathy toward deadlines and compromise. These characteristics make it more difficult to provide useful assistance to medical entrepreneurs.

Calgary's Projects Program

The central feature of the Projects Program offered through the Entrepreneurship and Innovation Area in the Faculty of Management at the University of Calgary is its reliance on graduate (and undergraduate) student teams to complete specified assignments that will assist entrepreneurs in the formation of their businesses. The educational background of the students is both broad and interdisciplinary, including engineering, environmental design, management, law, the sciences, and other disciplines. As part of their assignments in specific courses, students typically work with mid-sized entrepreneurial companies, firms of 25 to 150 employees, in teams of three to five, and in return the companies pay a nominal fee to the Faculty of Management's Projects Office for administering the program. The intent has been to identify raw potential in individual entrepreneurs and/or projects, and then to facilitate their progress to commercialization. In the process, the expectation is that both students and clients will learn more about the process of venture creation and development.

Client referrals come from multiple sources, including University Technologies International (technology licensing), the Calgary Research and Development Authority (incubator program), and the Alberta Research Council (funding source). The Projects Program also employs these and other sources of support as potential additives to its resource mix. All of these supports are necessary to achieve the flexibility in information inputs, instructive processes, and prescriptive outputs that the project approach requires. Students also have access to the advice of management professors, mentors from local businesses, and their own classmates. Strategic alliances with the Faculties of Law, Science, and Engineering have also been developed, allowing students to benefit from the assistance of persons trained or training in those disciplines. This opportunity provides unique cross-disciplinary experiences for the students – a distinction from most other student project programs – and more substantial technical, legal, and business information for entrepreneurial clients.

Experience suggests that the success of a project depends on the willingness of the entrepreneur to commit time to work with the student teams. The projects that appear to add the most value to both student education and entrepreneurial companies are those that address narrowly defined and specific questions. In addition, when expectations and deadlines are clearly defined and documented early in the process, there is a greater chance of project success. To secure greater commitment and interest, projects are screened prior to use in the classroom and are carefully matched with the educational objectives of a course. Once assigned to a course, students are generally permitted some discretion in picking their projects from the approved sets of clients.

Other aspects of project implementation which facilitate success include client follow-up, debriefing, and careful delineation of projects (e.g., opportunity analysis, market research and planning, and strategic, financial, and business planning).

Matching the Program and Medical Entrepreneurs

We have noted a number of significant differences between medical/academic entrepreneurs, and managers/businesspeople. We have also discussed the potential for conflict implicit in these differences. The Projects program provides an opportunity to bridge the gap among the stakeholders in the medical technology commercialization process.

The challenges in matching the capabilities of the students and program with the needs of medical entrepreneurs and the marketplace are many. The most important consideration is to ensure that in matching the medical entrepreneur with the marketplace, circumstances that lead to learning are created. The learning that occurs should highlight the need for all stakeholders in the process to make concessions if success is to be achieved. More so than in any other area of technology innovation, boundary spanners are a valued commodity in the area of medical technology. The challenges of friction between managers and individuals with technological orientations, and the failure to recognize management-technology systems interdependence apparent in the overall domain of technology management, are evident in medical technology as well.

The Projects Office is reasonably effective at making a successful match among clients, courses, and students. Much of the challenge encountered by the office results from dysfunctional interaction among participants and the politics of the various organizations involved. The matching problem in the area of medical technology ventures is complicated by the regulatory environment in which medical projects operate, and the social issues that pervade this environment.

Benefits of the Projects Program to Medical Entrepreneurs

As noted above, it is our contention that the Projects Program should yield parallel benefits for those involved in medical entrepreneurship and those engaged in other areas of enterprise. Given their comparative technological advantage and managerial disadvantage, medical entrepreneurs should also profit from interaction with students in a project. The imbalance between medical entrepreneurs' knowledge of science and their knowledge of business should be redressed by continuous, gradual immersion in the marketplace. Since primary contact would be with student teams, the students would serve as intermediaries and partially insulate, or "buffer," the less experienced entrepreneur from the exigencies of the marketplace. Since students would be serving as consultants and only making recommendations, medical entrepreneurs should not perceive them as threatening or likely to usurp decision-making power. Given their aforementioned sensitivity to intervention by non-technical managerial types, medical entrepreneurs could be expected to display a higher level of tolerance toward students because they perceive a greater degree of control over their situation.

Aside from the advice received, and the potentially increased willingness of medical entrepreneurs to take the advice of students, another potential benefit of the program is the development of medical entrepreneurs' ability to recognize the value of the input from intermediaries in dealing with the vagaries of the marketplace. Continued interaction with the students, and the realization of the validity of many of their recommendations, should enhance both the team's credibility and the business expertise of the entrepreneur. As entrepreneurs gain a better understanding of the nature of the marketplace, they should experience decreased resistance to interaction with the consultants. Recognition of the short-term pressures and operational constraints of business should further facilitate acceptance of the consultants' recommendations.

Medical entrepreneurs might also benefit from projects because of the increasingly complex legal and regulatory environment in which they function. With ever-escalating potential for legal liability and the requirements for economic justification for medical products, higher standards of due diligence are being imposed. The necessity for legal, economic, and managerial expertise in commercializing medical, and other closely regulated technologies is therefore frequently greater than in less constrained entrepreneurial settings. Thus, medical entrepreneurs may potentially gain even more benefits from the interventions available through the student project milieu.

Through the buffered discovery mechanism of working with student teams, medical entrepreneurs should gain an appreciation of how perfectionism, ignoring deadlines, failure to compromise, and rejection of the importance of profitability in their entrepreneurial pursuits are dysfunctional. The distinctive but overlapping roles of research and commercialization should also become evident, making the inevitable conflict between these two pursuits easier to reconcile and manage. Thus, the potential for reducing or resolving difficulties between entrepreneurs and other players is evident.

In economic terms, the potential benefits available from the Projects program offered by the University of Calgary are large. Chrisman's (1997) study of past clients who used one of the various services offered by the Entrepreneurship and Innovation Area from 1990 to 1994 indicated that the average value of the Projects Program was estimated to be $4,500, at a cost to the entrepreneurs of approximately $500.

A Preliminary Assessment of Calgary's Projects Program

Given the theoretical possibility of benefits to medical entrepreneurs from projects conducted by students, it is appropriate to examine the actual benefits that have accrued to medical entrepreneurs from the University of Calgary's Project Program.

In the spring of 1997, a questionnaire was sent to 144 entrepreneurs who received assistance from the Projects Program offered by the Entrepreneurship and Innovation Area of the Faculty of Management of the University of Calgary during the previous two semesters. Of the 40 clients who returned the survey, 10 were associated with medically related organizations (100% response rate), and 30 with a diverse range of non-medical firms (22% response rate). Respondents were asked to rate the quality of the program and to assess the benefits received and their satisfaction with the program on a number of dimensions, using a five-point Likert scale (1 - strongly disagree or very unsatisfied to 5 - strongly agree or very satisfied). The overall sample was wide-ranging – from software services and financial institutions to public sector organizations. The medical clients included dental, physician, home care and other medically related practices.

Table 1: Evaluation of Entrepreneurship and Innovation Project Programs' value by medical and non-medical entrepreneurs

Question Content	Medical (N=10)	Non-medical (N=30)
Project met process expectations	3.4	3.7
Project met financial expectations	4.2	4.1
Project met quality expectations	4.0	4.0
Students demonstrated competence	4.4	4.2
Students demonstrated professionalism	4.0	4.5
Project saved me time	3.6	4.0
Project saved me money	3.8	3.6
Project saved me effort	3.9	4.0
Project added value	3.6	3.9
Satisfaction with project	4.0	4.0
Satisfaction with students	3.9	4.2
Would do another project	3.9	4.1
Would recommend this service	4.3	4.4

No significant differences found between the two groups for any of the variables studied.

It is instructive to explore the responses of the medical organizations, both segregated and in comparison with the non-medical sample (see Table 1). Given the small sample, the responses should be interpreted with care. In their review of entrepreneurship education, Gorman, Hanlon, and King (1997) further note that empirical research on entrepreneurship education is exploratory, with most studies measuring key variables by self-report.

Overall, there were no significant differences found between the medical and non-medical groups for any of the measures employed. Thus, the global measures of client satisfaction with the project, the students, and the Faculty of Management Projects Office appeared to be similar for both groups. Furthermore, both groups were very interested in participating in another course-based project and even more likely to recommend this service to others. Other general measures that yielded similar responses from both groups included the meeting of client expectations with respect to process, finances, and quality, the reasonableness of administrative fees charged, the time, money, and effort saved through the project and ancillary services, and the overall value of that project and services.

Nevertheless, and worth mentioning due to the limitations of making inferences from a cross-sectional study with a small sample, there were some areas where differences between the groups approached significance. The medical sample felt that it was less adequately informed as to the nature of the course prior to the project than did the non-medical group. Medical entrepreneurs were also somewhat less pleased with the adequacy of contact with students and with their level of professionalism.

Still, given the slightly more critical response of the medical group, their item scores nonetheless averaged well above the scale midpoint of 3.0. This paints a positive portrait of the experience of both medical and non-medical entrepreneurs participating in the project-based program. This is especially true when considered alongside the positive responses from both groups in respect to student competence and the usefulness of project reports in providing new and valuable information and concepts, operational recommendations, enhanced decision-making, and a small impact on short-term sales and business practices.

Case Studies

The above analysis provides some preliminary information on medical entrepreneurs' perceptions of the benefits of the Projects Program. While such information is useful in support of the contentions made earlier in this article, it lacks the depth necessary to gain a full appreciation of the assistance that student projects can provide in the development of a venture. The remainder of this section will, therefore, present two short case examples of the value of student projects in more constructive terms.

Neurokinetics

Neurokinetics was a firm founded to develop and market a "bionic glove," which provides basic grip function to medical patients whose muscles have atrophied because of neurological damage (Graham and Graham 1995a, 1995b). The grip glove is based on a scientist's research into functional electrical stimulation technology (FES).

Prior to obtaining assistance from the Projects Program, the principal had developed an earlier application, also based on FES. In the prior case, he had published papers on the application without patent protection, thereby losing the exclusive rights to the product he had developed. The scientist did not want to repeat this experience with the new glove. He also found that the new company he had started,

Neurokinetics, had consumed far more time and effort than he had anticipated. It was difficult for him to balance his research efforts with his commercialization project.

After receiving funding from the Alberta Heritage Foundation for Medical Research, he solicited the assistance of three MBA students in the Entrepreneurship and Innovation program. The MBA students conducted market surveys and recommended a lean organizational structure in the early stages of commercialization. They also developed a business plan and provided information on the industry, target markets, lead times, competition and regulatory restrictions. In launching Neurokinetics, the principal was guided by the market research of the students in making initial product development decisions.

Following clinical trials and subsequent product development, seed investment was secured through a number of sources. Even with the resources the scientist had obtained, he still found his own level of involvement with the glove's commercialization unacceptably high. He wanted to continue development of the device, but did not want his teaching, research, or reputation to suffer.

Neurokinetics began operation in 1994 and had secured much of the resources required to complete its product development. One year later, it sought support for final commercial development of the glove. At that point, a promoter of spin-off companies recommended that Neurokinetics pool its resources with a second spin-off company, Bio Motion, since both firms based their products on FES.

The new company, NeuroMotion, obtained the necessary resources, and the media ("NeuroMotion," *Calgary Herald* 1999) reported its continued success. NeuroMotion's WalkAide, a medical aid that permits stroke and spinal cord patients to walk more comfortably, is undergoing extensive testing and promises to fill a critical niche in the rehabilitation market.

Tenet Medical Engineering Inc.

Student project teams have also played a crucial role in the success of Tenet Medical Engineering Inc. Founded by medical engineers at the University of Calgary, Tenet designs and develops patient positioning aids for surgery, surgical instruments and equipment, and custom patient positioning devices and surgical/hospital equipment (Alberta Economic Development, 1997). Entrepreneurship and Innovation students provided market research, market and product scope advice, and assistance in developing a business plan. Following the adoption of these recommendations, Tenet successfully initiated the early stages of its final commercialization/launch and expects to market its products in Europe and other foreign markets shortly (Alberta Economic Development 1997).

Conclusions

Although the evidence presented in this paper is only preliminary, there appears to be both a conceptual and an empirical rationale for concluding that the Projects Program offered at the University of Calgary is valuable to medical entrepreneurs, as well as to the larger enterprise community. The matching process, together with the great degree of flexibility built into the program, appears to fulfill the requirements of, and minimize conflict with, medical entrepreneurs. An effective balance seems to have been struck between the commercialization and educational objectives of the program, and its adoption and use by medical entrepreneurs. Review of the limited data available suggests that the program's value may be generalizable from overall entrepreneurship to the specific subset of medical entrepreneurship.

The value added by the program (including baseline values for the entrepreneurial development process), as well as other economic and scholarly benefits, has been briefly addressed and should be further elaborated as additional data is accumulated. Less quantitative, but nonetheless important, data on the expected intangible benefits to all stakeholders in the program should also be accumulated. It is crucial that discussion of appropriate criteria for program assessment proceed without delay (Gorman, Hanlon, and King 1997). Both internal and external validity of entrepreneurship research should be emphasized, as well as any methodological challenges that limit such validity. Enhanced program evaluations will aid in the comparison of different entrepreneurship programs, and the determination of the appropriate use of each.

An examination of a number of suggestions for unfreezing the process of commercialization of medical technology, in the context of change management, should also be considered. Unfreezing the time constraints of the program (i.e., freeing it from the academic calendar, instruction in blocks, etc.), and the nature of publications resulting from its operation (increased publication of cases, critical incidents and descriptive/prescriptive material on the entrepreneurial process) might bear discussion. The latter alternative would provide guidance and role models for practitioners, as well as aiding in the demystification of how medical technology can be commercialized and its presence broadened (by extending its presence into Medicine and other areas where it is currently less visible).

In conclusion, the preliminary evidence presented in the paper suggest that the Projects Program of the University of Calgary may assist medical entrepreneurs in the process of commercializing the technological discoveries and inventions emanating from their research. Further development and experimentation with the concept is justified. Most important, however, is the insight that the Calgary experiment could provide to public policy makers in western Canada who seek to encourage and enhance medical entrepreneurship on a wider basis.

References

Alberta Economic Development (1997). Tenet Medical Engineering Inc. *The Tech Files,* www.gov.ab.ca/edt/techfiles

Chrisman, J.J. (1994). *Economic benefits provided to the Province of Alberta by the Faculty of the University of Calgary.* Calgary: University of Calgary.

Chrisman, J.J. (1997). Program evaluation and the venture development program at the University of Calgary: A research note. *Entrepreneurship Theory and Practice* 22(1): 59-73.

Corsten, H. (1987). Technology transfer from universities to small and medium sized enterprises – an empirical survey from the standpoint of such enterprises. *Technovation* 6: 57-68.

Drucker, P.F. (1985). *Innovation and Entrepreneurship.* New York: Harper and Row.

Foxall, G., and Johnston, B. (1987). Strategies of user-initiated product innovation. *Technovation* 6: 169-187.

Gorman, G., Hanlon, D., and King, W. (1997). Some research perspectives on entrepreneurship education, enterprise education and education for small business management: a ten-year literature review. *International Small Business Journal* 15(3): 56-78.

Graham, B., and Graham, J. (1995a). Neurokinetics Inc. Unpublished case. *Alberta Heritage Foundation for Medical Research.*

Graham, B., and Graham, J. (1995b). NeuroMotion case study. Unpublished case. *Alberta Heritage Foundation for Medical Research.*

Huefner, J. C., and Hunt, H.K. (1994). Broadening the concept of entrepreneurship: Comparing business and consumer entrepreneurs. *Entrepreneurship Theory and Practice* 18(3): 61-78.

NeuroMotion markets device to enhance mobility, (1999, February 25). *Calgary Herald*: A27.

Quintas, P., Wield, D. and Massey, D. (1992). Academic-industry links and innovation: Questioning the Science Park Model. *Technovation* 12: 161-76.

Samsom, K.A., and Gurdon, M.A. (1993). University scientists as entrepreneurs: A Special Case of Technology Transfer and High-Tech Ventures. *Technovation* 13: 63-72.

IV

Innovation, Entrepreneurship, and Management in Family Businesses

Chapter 12

Investing with Families in Business: A Venture Capitalist's View

R.A. Slator, edited by Jess H. Chua

Abstract

This edited speech summarizes the experiences of a venture capitalist in investing in both family-owned enterprises and enterprises where a family has not been one of the critical elements of the organization. The author identifies four ingredients for a successful business, how they tie into family-owned enterprises, and whether family involvement enhances or hinders an organization's ability to achieve success. The author also highlights some of the key issues that an independent investor should consider when evaluating a family-owned enterprise seeking financing. The author concludes that, from an investor's point of view, the major emphasis remains on the merits on the business deal itself, but that family issues can affect this judgment.

Introduction

Twenty percent of our investments over the past eleven years have been in family-owned enterprises. This family involvement has been uniquely western Canadian because none of our family-owned businesses are in eastern Canada or the United States. Instead, they have all been located in Alberta. This was not by design, so the investment mix suggests that family-owned enterprises, on a per capita basis, may be more prevalent in western Canada.

Our experience with family-owned enterprises indicates a success ratio of about 50%. Significant alterations in the family compact had to be made, however, in 50% of our successful ventures. Therefore, only about 25% of the investments in family-owned enterprises proved to be successful without significant change.

This paper is a summary of the learning from our experience with investing in both family-controlled enterprises and enterprises where a family has not been one of the critical elements of the organization. The paper is organized as follows. The next section outlines what venture capitalists consider to be the ingredients for a successful business, regardless of whether they are family businesses. This is followed by a discussion of the special problems encountered when venture capitalists invest in family firms. Conclusions and recommendations to family firms are made in the final section.

It must be pointed out the opinions expressed here are mostly representative of those held by venture capitalists. Venture capitalists perform a particular function in the capital market. Investors that fill other roles in the capital market may hold views that are significantly different.

Ingredients for a Successful Business

Venture capitalists take enormous risks when investing in the early stage firm. For them, success is realized only if the enterprise grows rapidly to a significant size and yields very high returns to compensate for the risks taken. To attain this level of success in the current competitive market, where goods, services, capital, and people move across ever-lowering barriers, a business must have four basic ingredients. These are: people leadership, market leadership, a global perspective, and appropriate financing. These same ingredients are required regardless of whether the business is family-controlled. Each of these ingredients is discussed in detail below.

People leadership

By far the most critical ingredient for success is people leadership. Here, leadership does not just refer to the leadership of the chief executive officer. Rather, it is about leadership throughout the organization, from the board of directors, management team, and various department heads, to every employee of the company. The CEO may set the tone, but he or she cannot be in all places at all times. Leadership and furtherance of the company's goals and objectives must come from within all levels of the organization, and the climate must be set where that process should best occur.

Volumes have been written about leadership, and it is not the purpose here to discuss the subject in general. Instead, we concentrate on the specific qualities that a venture capitalist considers to be of critical importance.

First and foremost among these qualities is integrity. If the highest degree of integrity is absent, none of the other attributes matter at all. Next are the people skills of the management team. They must be able to inspire confidence and belief in the task at hand, managing the very difficult process of building a business. Following this, the people in the business must, as a group, possess all of the business skills necessary to build a successful business in the particular chosen industry. Finally, the people must work as a team.

None of these attributes are easy to assess. The skills are mainly learned through experience. In the end, one often has to rely only on gut feelings.

Market leadership

The second key ingredient for success is contained in the company's strategic approach to the market. The company must have a product, service, or technology that is accepted by the market. It must execute all of its regular and daily tasks well. But, this is not enough. In addition, the firm must attempt to achieve market domination or at least a market leadership position. Market dominance or leadership can be achieved by following various approaches, which are best illustrated with examples.

The first example of market leadership involves cost leadership. The company is an Edmonton-based manufacturer of sodium chlorate, an environmentally sound replacement for chlorine, used to bleach pulp and paper. The two primary inputs in manufacturing sodium chlorate are brine and electricity. To save on transportation cost, the company located its plant atop a 10,000-year-supply salt well. In combination with the fact that the Province of Alberta has the second lowest electricity cost in North America, the company is unbeatable as a low cost producer. In the commodity business, the best market strategy is to be the low cost producer. The company can land sodium chlorate in New Orleans more cheaply than a competitor located 100 miles away in Baton Rouge. The cost advantage of this particular firm is so great that its output for the first five years was sold before the ceremonial sod-turning for construction. Three years later, the firm was already considering doubling its plant capacity, to be financed through pre-selling the additional output prior to construction.

The second example of market leadership is achieved through market coverage. The company is a sporting goods retailer based in Calgary. It was a reasonably successful $20 million family enterprise when we invested in it. Through acquisition of four other sporting goods chains, it became the dominant sporting goods retailer in Canada with sales of $350 million after three years. The company actually ranks number four in North America. This dominant position provided the firm with a buying power that cannot be exceeded by any other group and that will ensure that it can withstand the onslaught of U.S. category killers moving into Canada.

The third example of market leadership is the creation of a unique niche that the company can dominate. Three brothers, all under 30, started this company in Edmonton from scratch. Their specialty was manufacturing flame- and chemical-resistant clothing for industrial use. They built their company from zero to $30 million in sales in three years and will, in all likelihood, exceed $100 million dollars in another three years. Their market is global, and they have been able to achieve this market dominance because of an incredible gap that existed within their industry. There existed no standards for this form of protective clothing. The three young men got together with design groups, government regulatory bodies, and major fiber suppliers, to design and create the standards that their entire industry worldwide must now follow. Clothing manufacturing is supposedly done most cost-effectively in Asia, but creative minds and the entrepreneurial talent can overcome any cost disadvantage.

The fourth and final example of market leadership is world-leading quality. The company is the world's top quality producer of protocol testers for the telecommunications industry. Nippon Telephone and Telegraph (NTT) sent their vice-president of quality control and eight staff to North America to find a supplier for protocol testers. They found AT&T, IT&T, and this company in Edmonton. The company's product was the only one that passed NTT's quality control tests. The company was sold to Hewlett-Packard for a huge profit to its investors.

Market dominance tends to be associated with giant companies, but three of these four examples have sales of under $50 million. They are not large but will continue to be the supplier of choice because of their market leadership.

Global perspective

The third ingredient for success is a global perspective. In the global economy, borders no longer exist except in the minds of politicians. Any business that believes in borders will not be in business long. Even a niche supplier must now attack the market on a scale much broader than local or regional. The best defence is a good offence, especially in business today. Any business that is successful in local or regional markets must act very quickly to broaden its base because competitors will get wind of the success and invade its market. These competitors will not be from around the corner as in the "good old days." They will be from Asia, Europe, and Latin America.

Globalization can be done by virtually any size of company. There was a time when only the General Electrics and the General Motors of this world could operate on a global basis, but the technological advances made in communication and transportation have opened the global market to the smaller companies as well.

The companies used as examples in the previous section are all examples of global perspective although their sales are under $50 million. The chemical company can deliver its products to its markets at a cost lower than anyone else in the world. The sporting goods retailer buys its goods at the lowest cost that its global suppliers charge. The safety clothing manufacturer defines what quality is in that niche and is in a better position than anybody in the world to take advantage of it. The protocol tester developer sells products that surpass the highest standards set in the global market.

Appropriate financing

The final ingredient for success is financing. In order to develop the other three key ingredients, people leadership, market leadership, and globalization, the equity base must be able to support aggressive growth. That aggressive growth can only be achieved by sharing the company with other investors, because a high growth business cannot generate enough free cash flow to feed the growth. The entrepreneur must believe that it is better to have 40% of a successful $50 million company than have 100% of a $5 million company that is just getting by.

An appropriate equity base and sound equity partners will bring many competitive advantages that others in the industry will not enjoy. Banks are more likely to lend working capital to help finance expansion plans, and terms and conditions of financing will be much more attractive. Nothing frustrates an entrepreneur more than being unable to take advantage of market opportunities because of working capital shortage.

Suppliers will be much more accommodating to a well-financed organization by providing better terms if the company is well financed for extensive growth. Customers will be more likely to sign long-term contracts if they know that the business will be around in the future. Finally, appropriate financing will help the company to attract the best employees and keep them loyal.

High involvement investors, such as angels or venture capitalists, can provide access to international people networks, opening doors to potential markets, financiers, customers and board members. These high involvement equity partners can share in the inevitable pains and work of management to achieve the success that both so very much desire.

In summary, venture capitalists are interested only in businesses that have potential for global market leadership and, thus, a sizeable return on their investment. In order to achieve this objective, the firm must have the people capable of, and committed to, realizing the potential. These businesses are the ones that can benefit most from the equity base and close working relationship supplied by the venture capitalists.

Issues Specific to the Family-Controlled Business

In the previous section, we suggested that, for a business to have the potential to meet the venture capitalist's definition of success, it must have four key ingredients: people leadership, market leadership, a global perspective, and a solid financial base. These ingredients are essential for all businesses, whether family-controlled or not. For the family firm, however, whether these ingredients are developed and enhanced or destroyed depends critically on the attitudes of family members and their working relationships.

Many venture capital funds in the United States have publicly stated that they will not invest in a family business. Our successes were very positive involvements, both from a financial and an experiential point of view. In addition, we learned through the process and believe that we would not repeat our mistakes. In the discussions below, we review what venture capitalists consider to be issues specific to the family-controlled firm.

Market leadership versus lifestyle

Both in Canada and the United States, most family-owned enterprises are essentially lifestyle businesses where the objective is to produce a good living for the entrepreneur and his or her family. These businesses may make good and valid contributions to the national economy, but because they are more local, as opposed to global, in scope they are not really of much interest to the venture capitalist. Businesses devoted to maintaining a good living for a family will not be committed to market leadership. In addition, they are unlikely to desire a constructive working relationship with a venture capitalist, typically seeing advice from such a person as interference and requests for information as invasion of privacy.

Family member rivalry

Sibling rivalry and parent-child rivalry can tear a company asunder. Many family-dominated public companies have been split apart because of family members' irreconcilable views or the fact that they just do not like each other anymore. The number of private companies that split apart for the same reasons is unknown but likely larger.

Anyone with experience in mediating these disputes understands the reluctance of police officers to be involved in family fights. The outside investor usually ends up being the bad guy because no one else will do what has to be done. For example, in one situation, an investor had to ask the father of a family enterprise to resign because the rest of the family, who held the majority of the shares, were unable or unwilling to do so. The patriarch simply did not have the energy or the desire to lead the company to the next stage of growth, but no one was willing to tell him. In another situation, investors had to gain the support of a father, a brother, and a sister to oust their sibling as the CEO of the company, even when failure to do so would have doomed the company to certain bankruptcy. Consequently, it is much harder to achieve people leadership in the family firm because family member rivalry will inevitably occur and detract from the working relationships, as conflicting parties in the family context cannot be easily separated.

From the outside investor's point of view, he or she must be comfortable with the people in the firm and believe that he or she can work constructively with them. This is, interestingly enough, very important because investors have found, over the years, that life is too short to have to deal with people that one just does not like. Constant fighting among family members will seriously detract from a family firm's attractiveness to the investor. Without outside equity, it is very rare for a family firm to achieve market leadership on a sound financial footing.

Nepotism

A separate problem with people leadership in a family-owned enterprise is the perception (and quite often the reality) that family members are treated differently when it comes time for promotion, salary increase, or perks. This perception can cause good non-family people to leave the enterprise and make it difficult for the company to attract quality people.

Our worst experience was with a company that had six family members in various management positions. We were unaware of this situation because in-laws who had different last names occupied four of the six positions. The company went bankrupt and was a loss for both the investors and the family.

Clarity of leadership

Family-owned firms have the tendency to delay sorting out the leadership issue. Clearly, if no one knows or agrees who is/are the leader(s), there can be no people leadership. As an example of what could go wrong, take the business started by three brothers, with one of them as the CEO. To avoid dealing with the issue of who was worth more to the company, the brothers decided that they should receive exactly the same compensation, even though their titles and responsibilities and authority were otherwise defined. There were, at many times, questions of leadership, sometimes causing problems within the management group. In another company, again with three brothers, the middle brother was the CEO. The older brother had seniority in the family hierarchy and expected that to carry through to the business. The other shareholders and the other brothers did not agree. The older brother had to leave the company because he could not abide reporting to his younger sibling.

Conclusion and Recommendations for Family Firms

In conclusion, a business, whether family-owned or not, will be successful in the eyes of a venture capitalist only if the following requisites exist in the business. There must be a product, service, or technology that gives the firm the potential to achieve global market leadership. There must be a quality management team capable of, and committed to, achieving the potential. The product, service or technology must enable the establishment of an enterprise significant enough to make building the business worthwhile for both the entrepreneur and the investor.

If the business is family-controlled, there are further difficulties that must be overcome. Before even starting, there is already one strike against the family firm just because it is a family enterprise. As discussed, most venture capitalists view extensive family involvement as a negative.

Accordingly, the family firm has to sell harder to overcome the scepticism. The family-owned firm must convince the investor that it intends to pursue aggressive growth rather than a lifestyle chosen by the family. It must convince the investor that the firm will be a major enterprise.

Family-owned enterprises must not only make every effort to reward and promote on the basis of merit, but must be seen or perceived to be doing so on a regular and consistent basis. This must hold true for both family and non-family members. As the saying goes: "Fair is not equal."

Family members must convince the investor that they have the leadership question sorted out within the family group and that family dynamics will not act as an impediment to achieving success. Investors will look for all sorts of hints and keys that this issue has been dealt with and that all members of the family group have bought in to the solution.

If these special problems can be overcome, the family firm will have a greater cohesiveness within the share ownership of the organization. The group will be bonded together by their shared desire to achieve success as a corporation and, even more importantly, they will be driven by the potential success of the family name.

Chapter 13

Your Old Men Will Dream Dreams, Your Young Men Will See Visions: A Conceptualization of Innovation in Family Firms

Reginald A. Litz, and Robert F. Kleysen

Abstract

Despite significant advances, a conspicuous gap remains in family business research. Specifically, is there anything distinctively different about the practice of innovation in family firms as compared to non-family firms? After reviewing a selection of literature on innovation and family business, we offer coarse- and fine-grained conceptualizations of intergenerational innovation in family enterprise, together with supporting anecdotal examples. Given the fine-grained distinctions inherent in our resulting definition of family firm innovation, we move on to in-depth study of a family involved in the innovative activity of group jazz improvisation. After offering our analysis of the core dynamics apparent in this family setting, we conclude our paper with a research agenda for future work on family firm innovation.

Rome falling to the barbarians, an old family firm going into bankruptcy, and a government agency quietly strangling in its own red tape have more in common that one might suppose.

- *J. W. Gardner (1964, 3)*

If you love your children, send them on their travels.

- *Arabian proverb*

Introduction

What is the relationship between the presence of family within a business enterprise and its capacity for, and practice of, innovation? While family firm research has made significant inroads in recent years (Dyer and Sánchez, 1998), this question has received limited attention. However, the issue of how intra-organizational familial involvement affects organizational innovation is of critical importance to practitioner and researcher alike. Consider the often cited statistics – namely that 70%

of family firms fail to successfully pass the baton to the second generation, and that only about 10% successfully pass it on to the third generation (Kets de Vries 1993). At a minimum, this statistic suggests that the typical family firm lives a precarious existence fraught with challenge.

Why is this so? Why do so many family firms fail? As with most phenomena, the outcome is overdetermined; issues of parental relinquishment, successor incompetence, sibling rivalry, and industry evolution all play a part. However, alongside these explanations is another, all too often overlooked, variable: a failure to innovate. As Gardner's opening quotation suggests, the ongoing challenge of decline is faced by every organization. However, what is intriguing in Gardner's statement is his insertion of the adjective *family* before the noun *firm*. While this might indicate a stereotypical assumption, namely that "when family and firms are interrelated, a less effective business enterprise generally results" (Donnelly 1964, 55), it might also signal a more profound observation – that intra-organizational family involvement spells eventual organizational dysfunction.

More than one reason for such an assertion comes readily to mind. First, at times family firms appear unable to reckon with the demanding realities of the marketplace, due, in part, to the melding of firm strategy and family identity. A second reason is also apparent: family patriarchs, exempt from the discipline of formally imposed succession practices, are too often permitted to ossify in senior executive positions (Sonnenfeld 1988).

However, some family firms do manage to avoid these dysfunctional patterns. Consider Goldwasser's (1986) study of five excellently managed family businesses: Hallmark, Noxell, Marriott, H & R Block, and Johnson Wax. Firms such as these and others pursue novel product and service opportunities and lead their respective industries in creatively attaining and sustaining competitive advantage. How do these family firms defy the ominous mortality rates cited earlier? Furthermore, how do these family firms not just survive, but more specifically thrive, as they lead their respective industry cohorts in innovative initiatives?

In this paper, we consider these questions. We begin by reviewing work on intra-organizational innovation. After considering a selection of seminal pieces, we assert that work to date has largely excluded a variable significant to the vast majority of business enterprises: the intra-organizational presence of family. After reviewing existing research on family business, we proceed to explore the interaction of these two constructs, namely, innovation and family business.

Our integrative conceptualization begins with a coarse-grained typology of intergenerational innovative behaviors. Utilizing several family business biographies, we offer supporting examples for each of the proposed types. We then explore the underlying issues suggested by our coarse-grained typology by advancing a finer-grained and more exhaustive conceptualization and definition of family business innovation. Building on our definition, we take an in-depth look at one family active in the innovative arena of jazz improvisation. We conclude by noting several important definitional and methodological issues in our proposed framework, and offer several suggestions for further research on family firm innovation.

Theoretical Context

The importance of innovation

Ever since Schumpeter's (1942) emphasis on the key role played by innovation in the progress of organizations and societies, both business practitioners and academics have regarded innovation as a central managerial concern. Quinn (1988) argued that innovation management is one of only a few strategies available to meet the increasing challenges of international competition. Academic interest in innovation is reflected in attention devoted to the topic as evidenced by special issues of leading research outlets, such as *Administrative Science Quarterly* (Tushman and Nelson 1990), *Strategic Management Journal* (Guth and Ginsburg 1990), and the *Academy of Management Journal* (Drazin and Schoonhoven 1996) devoting entire issues to its study. Concern for innovation amongst practitioners is evidenced by longitudinal work carried out by Van de Ven and Associates (Van de Ven, Angle, and Poole 1989) sponsored by several organizations, including the U.S. Navy, 3M, IBM, Honeywell, and others. Contrary to Kimberly's (1981) prediction that a more skeptical view of innovation likely was going to emerge in the 1980's and 90's, interest in innovation has ballooned.

Defining innovation

But what does this ubiquitous and positively connoted term, innovation, really mean? While many definitions have been advanced (Zaltman, Duncan, and Holbek 1973; Rogers 1983; Drucker 1985; Kanter 1988), perhaps one of the more integrative and useful ones was supplied by West and Farr (1989). They define innovation as the intentional introduction and application within a role, group or organization of ideas, processes, products or procedures, new to the relevant unit of adoption, designed to significantly benefit role performance, the group, the organization or the wider society (1989, 16).

For something to be labeled as an innovation, it must possess novelty, have a tangible quality, be the result of intentional rather than accidental action, be aimed at producing benefit, and be distinguished from routine change (King and Anderson 1995). To further demarcate the definitional parameters of innovation vis-à-vis related terms, creativity and entrepreneurship are assumed to be elements of innovation processes (Amabile 1988; Anderson and King 1993; Kanter 1988), while innovation processes are seen as part of the larger heading of organizational and cultural change (King and Anderson 1995).

In addition to the work aimed at defining innovation, several basic conceptualizations exist in the literature. Some researchers define various stages of innovation (Zaltman, Duncan, and Holbek 1973; Rogers 1983; Amabile 1988; Damanpour 1991; Anderson and King 1991), with such stages ranging from initial opportunity recognition and idea generation through to adoption and diffusion. Intra-organizational innovation research has also stratified by the level at which innovation occurs, including the individual (Kirton 1976; West and Farr 1989), the work group (Anderson 1992; West 1990), the organization (Zaltman, Duncan, and Holbek 1973; Staw 1990), and the industry/society (Rogers 1983). However, one mid-level grouping not explicitly considered is the family unit. At the same time, most authors also recognize the interactive quality of innovation between levels: as Kanter writes, "micro-processes [related to innovation] are...stimulated, facilitated, and enhanced, or the opposite, by a set of macro-level conditions" (1988,169).

Antecedents to innovation

Research has also explored antecedents to innovation by considering the different organizational levels involved in the process of initiation and implementation. At the individual level, for instance, divergent thinking (Guilford 1963), a propensity for risk-taking (Michael 1979; Glassman 1986), cognitive style (Kirton 1976; Sadler-Smith and Badger 1998), intrinsic motivation, domain-relevant skills, and creativity skills (Amabile 1983, 1988), political prowess (Maute and Locander 1994; Kanter 1983), and self-confidence, autonomy, and openness (West and Farr 1989; Angle 1989; Kanter 1988) have all been conjectured to lead to innovation.

At the group level, work on team climate (West 1990; Anderson and West 1998) as an antecedent of innovation has become increasingly important. The Team Climate Inventory (Anderson and West, 1998) measures four factors that facilitate innovation, including vision, participative safety, task orientation, and support for innovation. Other elements such as group composition, cohesiveness, development, social construction, and interpersonal interactions have also been discussed (Anderson 1992).

Finally, at the organizational level, Damanpour (1991) found centralization to be negatively associated with innovation, while specialization, functional differentiation, professionalism, managerial attitudes toward change, technical knowledge resources, administrative intensity, slack resources, and external and internal communication were all positively related. In other words, organic, rather than mechanistic (Morgan 1986) organizational structures tend to be more conducive to producing innovation.

Apparent limitations

While research on innovation has made important advances in the last few decades, several limitations also exist. First, debate over consistency of results persists; Downs and Mohr (1976) and Wolfe (1994) argue that findings have not been stable, while Damanpour (1991) provides evidence to the contrary. Second, though researchers have called for more process level theories of innovation (King 1990), work by Van de Ven, Angle, and Poole (1989) demonstrates that this is an extremely difficult and complicated task. Third, while significant advances have been made in understanding individual level innovative cognition and behavior, work at the group level is by comparison underdeveloped (Anderson 1992). This may, in part, explain the aforementioned gap in research on innovation in one of humankind's most central social groups, the family, and by extension, one of humankind's most enduring economic entities, the family enterprise. Before beginning to explore this unexplored territory, we first consider the conceptual and operational distinctives of the family firm.

The family firm: Origins, distinctives

Until recent times, family firms have been one of the most consistently overlooked organizational phenomena. The reasons are several and interrelated (Litz 1997). Since family firms are more frequently privately held (Dess and Robinson 1984), they are obligated to disclose comparatively less organizational and performance-related information than non-family firms. Compounding this is the interdependent nature of family and firm, resulting in a widespread hesitancy to disclose data on one, given its interpenetration by the other; in short, to talk about the business is to talk about the family. The lack of information results in researchers, intent on securing permanent tenure following probation, being hesitant to study that which is not readily researchable.

However, this trend has recently been reversed, beginning in large part with the 1988 launch of *Family Business Review*. As Dyer and Sánchez (1998) note in their recent review of the publication's first ten years, the past decade has witnessed a remarkable turn of events in family firm research, with a significant increase in conceptual, empirical and practitioner pieces. Looking beyond *Family Business Review*, research on family business has been featured in several leading outlets of entrepreneurship research, including the *Journal of Business Venturing* (exemplified by such efforts as Morris, Williams, Allen, and Avila, 1997), *Entrepreneurship Theory and Practice* (exemplified by Chua, Chrisman, and Sharma, 1999), and *Journal of Small Business Management* (exemplified by Covin, 1994). In some cases, entire issues (Hoy and Verser 1994) have been devoted to the topic of family business (for example, *Entrepreneurship: Theory and Practice*, Fall 1994 special issue).

However, a significant majority of work has quite predictably focused on the interpersonal and succession-related issues inherent in the merging of family and business systems. As Dyer and Sánchez note in their ten-year retrospective,

> the content of the literature on the field is likely where it should be at this stage. Because the field was developed to understand the interplay between family dynamics and business performance, it is not surprising that the most common topics are family relationships, succession, interpersonal relationships in the business, and the performance of family firms (1998, 291).

These authors, though, also go on to assert that "those working in the field need continued encouragement to provide an increasingly broader spectrum of interests" (1998, 291). In this paper we advocate that one such issue concerns how family firms facilitate intra-organizational innovation. To date there has been little work expressly devoted to the subject. While family firm sustainability has been addressed (Stafford, Duncan, Dane, and Winter 1999), the innovative dimension of firm performance (Chakravarthy 1986) has received at best indirect attention, with primary attention directed toward firm survival and sustainability. This is largely understandable, however. Given the horrendous mortality rates of family firms, and the emergence only recently of any significant body of family firm research, a predominant focus on simply achieving steady state existence is predictable. However, given the realities of today's rapidly evolving marketplace, a constricted view on only steady state existence falls short. What of the challenge of innovation, particularly when placed within a family firm context? To this question we now turn our attention.

Conceptualizing Innovation in Family Firms

Innovation in family firms: A coarse-grained conceptualization

What does it mean for a family firm to be a creative and innovative entity? In this section we address this question by offering a conceptualization of innovation in family firms. Building on the assumption that one of the central defining characteristics of family businesses is intergenerational involvement and that one of the core facets of creativity and innovation is envisioning and enacting that which is new, we take as a guiding inspiration a prophecy offered by the Old Testament sage, Joel. In the second chapter of his writings, he predicted a day when the people of Israel, both young and old, would be characterized by qualitatively superior insight, that is, when their "old men will dream dreams, [and their] young men will see visions" (Joel 2:28b, New American Standard Bible 1977). While Joel's original intent for this phrase was not, in all likelihood, to conceptualize intergenerational creativity, his phrasing nonetheless encapsulates the phenomenon's essential challenge: to facilitate the creative interaction of both founding and inheriting generations.

In order to assist in the discussion that follows, we offer a graphical depiction of Joel's statement in Figure 1. As the grid's coarse-grained depiction suggests, some families exhibit innovative initiatives across multiple generations (the northeast quadrant); however, others display such tendencies in either the older or younger generation (the southeast and northwest quadrants, respectively). In still others, innovative tendencies are present in neither generation (the southwest quadrant).

While our conceptualization is admittedly coarse-grained insofar as it fails to account for the possibility of either intra-generational creativity,

Figure 1: Coarse-grained conceptualization of innovation in family firms

	Non-innovative Older Generation	Innovative Older Generation
Innovative Younger Generation	Innovative younger generation (Visions without dreams)	Innovative older and younger generation (Dreams and visions)
Non-innovative Younger Generation	Neither innovative older nor younger generation (Neither dreams nor visions)	Innovative older generation (Dreams without visions)

such as between siblings Harrison and Wallace McCain at McCain Foods (Waldie and Jennison 1996), or the creative initiatives of non-family employees, such as Lee Iacocca at the Ford-family-controlled Ford Motor Corporation (Iacocca and Novak 1984), it nonetheless serves as a useful starting point. More specifically, it aids in providing a basic categorization scheme readily applicable to many family enterprises. In reviewing both published family firm biographies and anecdotal reports in the popular press, exemplars of each cell were readily identified. A brief review of a selection of these cases follows, together with our rationales for the linkages proposed.

Dreams and visions: Innovation as an intergenerational phenomenon

An example that comes readily to mind of a family firm where the creative impetus was shared intergenerationally is the computer colossus, IBM. At IBM, two generations of the Watson family made unique and complementary contributions to the company's strategic evolution (Watson 1990). While Tom Watson Sr. helped expand the firm's product mix beyond the simple technologies resident within the company at the time of his entry, his son, Tom Jr., undertook quantum transformations, leading the company into its dominant role in the mainframe computer industry.[1]

[1] The well publicized delay of IBM's late entry into the personal computer market would not likely be attributable to Tom Jr., as he had already retired from IBM in 1974. This failure to innovate by IBM would seem to be a result of other dynamics beyond our interest in *family* firm innovation.

Dreams without visions: The innovative parent

The southwest cell refers to those scenarios in which the elder generation had innovative tendencies that went unsupported by the generations that followed. An example of this form of interaction is the Sarnoff family at the Radio Corporation of America (RCA). Consider the experiences of father David and son Robert Sarnoff. The legacy of the elder Sarnoff was an impressive one:

> David Sarnoff was a salesman who compelled others to share in his dreams and make them come true by sheer force of personality. Not a scientist, he could induce scientists to invest their brainpower; not personally rich, he could induce capitalists to invest their money; not the company's majority shareholder – he owned only one-third on 1 percent of RCA stock in 1930 – he became its president the same year. By that time, RCA was mass-producing home radio sets, had established a broadcasting network (NBC), and was beginning to move into the chancy field of television. He predicted correctly that color TV would be a commercial success, although his concomitant prediction that this would lead to a new era of art appreciation was far off the mark. By the time David Sarnoff stepped down as company head in 1968, RCA had become a $3.3 billion a year company. (Alcorn 1982, 8)

On his retirement he was succeed by his son, Robert, who proceeded to attempt several new ventures, including an array of diverse and arguably unrelated acquisitions. However, the outcomes of Robert's initiatives paled in comparison to those of his father. As Alcorn notes,

> While his father's risky ventures had, by and large, led to company gold, Robert's similar efforts brought only company woe. He did abandon the computer business, into which his father had invested millions, because profits seemed a long way off, but he began to acquire a hodgepodge of companies that had to be merged somehow with an old-line electronics company. In short order, RCA was into car rentals (Hertz), publishing (Random House), real estate (Cushman and Wakefield), and carpets (Coronet Industries). Even David Sarnoff might have had difficulty turning this conglomeration into a unified business, and Robert had no success at all (Alcorn 1982, 9).

Visions without dreams: The innovative child

The third cell refers to those situations where the younger generation displayed innovative tendencies, unmatched by those of the elder generation. Owing to an apparent selection bias by family firm biographers toward documenting the histories of successful firms founded by dynamic, high-achieving entrepreneurs, examples of this state of interactions are less readily available. This bias notwithstanding, the history of Turner Broadcasting Systems is instructive. The company had its roots in Ted Turner rescuing his father's billboard company from bankruptcy (Thompson 1996). After re-establishing the firm's viability, he went on to diversify his holdings into radio and television broadcasting, sports franchising and a myriad of other holdings.

Neither dreams nor visions: Intergenerational stagnation

The final, and perhaps most perilous cell is the southwest, characterized by neither vision nor dream. Again, like the northwest quadrant, identifying exemplar firms is somewhat problematic. However, if we relax the assumption of the elder generation as founding and allow inheriting the firm, the interaction suggested by this cell is more easily visualized. For example, consider the case of the recently bankrupted Canadian retailer Eaton's. The firm, which commenced operations on December 8, 1869 as T. Eaton & Co. after Timothy Eaton purchased and renamed a dry goods and haberdashery business, was an unqualified success and rapidly expanded its operations to become Canada's premiere retailer.

However, founder Timothy never assumed the enterprise was assured of perpetual success. "All depends," he asserted shortly before his death, "on the integrity of those who are to follow after me" (McQueen 1998, 26). Timothy's words were prophetic; in the decades that followed, the trail of incompetent and indulgent Eaton descendants was almost uninterrupted, ending only with the organization's recent closing. During the industry evolution of the past decades, new players, such as Wal-Mart, and new retail customer value propositions, such as everyday-low-pricing, came to the fore. However, Eaton's stood firm in its traditions, in large part, because "no one had a vision for the future" (McQueen 1998, 118).

A finer-grained approach: Distinguishing autonomous and interactive innovation

While the logic represented in Figure 1 is readily understandable, at least one limitation is also apparent. This shortcoming is perhaps illustrated by looking more closely at the dynamics of one of the families mentioned previously, namely the Watson family at IBM. On closer examination it is apparent that in many ways, the inter-generational innovative initiatives at IBM were generation-specific. For example, Tom Watson Jr. is reported to have deliberately kept his father less than completely informed about the new product development efforts he championed; likewise, in an effort to buttress past accomplishments, personnel at IBM were explicitly prohibited from using the word "computer" while the senior Watson was still alive. Young (1998) recounts the intergenerational dynamics at work as the younger Watson clandestinely championed product development efforts that would eventually result in the company gaining valuable early experience in manufacturing computers:

> Watson Jr. knew this was the chance he had been looking for. Not only would he build the air defense machine to the specifications designed by the Project Whirlwind team, but he would assign a skunkworks group of IBM electrical engineers to work on the company's own big computer, one that could be sold to other government agencies or commercial accounts. By cloaking his ambition in the guise of his concern for national security, he intended to hide from his father that he was building a full-scale computer – what would be called the Defense Calculator. (The word "computer" was never used at IBM until after Watson Sr. died) (Young 1998, 42).

This anecdote suggests that the innovative practices in some family firms are only the independent efforts of individual family members, rather than the interdependent collaborations of the larger family unit. This distinction is important because it suggests that in the fullest sense family firm innovation must include more than simply the additive sum of individual family members' efforts. It must also include the possibility of innovations emanating from synergies between family members. This proposition has significant implications for conceptualizing innovation in family firms. At the very least, it requires that we distinguish between independent and interdependent efforts of family members.

Figure 2: Fine-grained conceptualization of innovation in family firms

	Non-innovative	Innovative
Interactive	Group of family members not involved in innovative effort	Group of family members involved in innovative effort
Autonomous	Individual family members not involved in independent effort	Individual family members involved in independent innovative effort

Conceptual and definitional implications

Building on these distinctions, between the innovative and the non-innovative, and the autonomous and the interactive, a working definition of family firm innovation can be advanced. Family firm innovation is defined as the intentional generation or introduction of novel processes and or products resulting from the innovative and interactive efforts of members of a family unit (depicted by the northeast cell in Figure 2). The latter element of interaction deserves special emphasis. Without it, the summed contribution of individual family members (as represented by the southeast quadrant) is virtually indistinguishable from the generic innovations of individual entrepreneurs, save family-based linkages between innovators, such as that between Tom Sr. and Jr. at IBM. This suggests that the crucial variable in family firm innovation is, therefore, the interaction between family members that results in collaborative innovative outcomes.

Beyond Definition to Description: A Case Study in Familial Innovation

Building on the distinction between innovations enacted by individuals related by family membership, and innovations enacted collectively as a family, a logical question emerges: temporarily leaving aside the institution of business, what would the practice of collaborative innovation by members of a family look like? To answer this question, we redirected our search beyond traditional organizational research subjects (in this case, family firms attempting to innovate), and focused instead on contexts characterized by high levels of collaborative creativity. Within such contexts, we then searched for one or more families that had successfully fostered high levels of intra-familial innovation. One context that especially lent itself to collaboration was jazz improvisation; a family unit active in that context was that of veteran jazz musician Dave Brubeck (Hall 1996). Before discussing the specific dynamics of the Brubeck family, we first review extant literature on distinctive characteristics of jazz improvisation in order to understand why it is inherently well suited as a backdrop for exploring the practice of family innovation.

Group jazz improvisation as a form of collective innovation

According to Bastien and Hostager, jazz is "an art form that is inventive and social [that] enables individual musicians to create new musical ideas in a collective context and, thereby, to achieve an inventive and integrated performance" (1988, 582). Compared to traditional performance, where "rehearsal is a means for working out an authoritative version of a musical innovation prior to group performance, [and] sheet music is a mechanism of constraint on innovation during performance, and... group performance largely consists of the reproduction of previously innovated musical ideas for an audience" (1988, 583), group jazz improvisation involves simultaneous creation during performance, without aid of either sheet music or rehearsal.

What is critical in group jazz improvisation is the interaction of three variables – information, communication, and attention, within the context of both musical and social structures. The underlying musical structure consists of the rudiments of music theory, that is, the "cognitively held rules for generating, selecting, and building upon new musical ideas, including rules for proper chords, chordal relationships, and chordal progressions" (1988, 586-87). In jazz improvisation the musical structure is comparatively loosely coupled (Weick 1976), as new musical ideas are neither entirely random nor predetermined, but rather are enacted within the context of a set of collectively held assumptions, including tempo, key, chordal structure and progression, phrasing, and embellishment.

By its very nature, group jazz improvisation also requires a supportive social context, that is, an appropriate set of behavioral norms and communicative codes. Behavioral norms are determined through the provision of nominal leadership (that is, the selection of the song, key, and starting tempo), solo leadership (in defining the complexity of a particular solo), shared leadership (in alternating solo leadership between players), and background support (playing chorus). Communicative codes involve not only "lexical items, or words and phrases of distinct meaning in the profession" (Bastien and Hostager

1988, 588), but also the more subtle non-verbal cues, such as eye contact, volume change, and body positioning. Jazz therefore "happens" as these two systems of musical competence and social relationship interact. Each makes a necessary, but not sufficient, contribution. The musical structure provides the melodies, harmonies, and rhythms heard, but only through the underlying social structure may the music be co-created.

A paradigmatic case study in family innovation: Introducing the Brubeck family

Given its inherently high need for collaborative creativity, jazz improvisation is extremely well suited for studying group innovation. As we searched for a family active in group jazz improvisation, one family repeatedly came to our attention – that of jazz pianist Dave Brubeck (Hall 1996). Brubeck was one of the individuals responsible for the post-Second World War emergence of jazz in the United States. His working relationships with fellow quartet members, including saxophonist Paul Desmond, bassist Eugene Wright, and drummer Joe Morello, resulted in dozens of recordings still in demand. Typical of full-time jazz musicians, he spent much time on the road with the quartet performing thousands of concerts in countless venues over their multi-decade association.

However, despite the arduous career-related demands, Dave also collaborated outside the jazz arena, most notably with wife Iola in the birth and nurture of five sons and one daughter. Four of the sons, Darius, Chris, Dan, and Matthew, decided independently of their father to pursue musical careers, specializing in keyboards, electric bass and trombone, drums, and cello, respectively. Darius currently serves as Director of the *New Center for Jazz and Popular Music* and Associate Professor in Jazz Studies at the University of Natal in Durban, South Africa. After pursuing formal musical studies, Chris started his own rock band, in which he played piano, eventually expanding his instrumental repertoire to include electric bass and trombone. More recently, he has also been involved in a wide variety of performance and composition projects, including choral works, musical comedy, and a concerto for trombone. Dan, who specializes in drums and percussion, has pursued a more rock-intensive direction, but nonetheless also remains active in jazz. Youngest brother Matthew is involved in *The Clubfoot Orchestra*, a ten-piece ensemble that specializes in the composition and performance of music for silent films. Not directly involved in music full-time, but no less members of the family, are second son Michael, who, because of a birth-related learning disability, was prevented from pursuing a career in music, and sister Catherine, who chose marriage and a career as wife and mother.

When a family does jazz: Two parents, six kids, five musicians

On several occasions Dave and his four sons have collaborated on musical projects. Outcomes include the 1974 album *Two Generations of Brubeck* and the 1997 *In Their Own Sweet Way*. How did these innovative projects happen? Drawing on available commentary, both from family members (as contained in the album notes of the 1997 album) and the documented reports of non-family observers (Hall 1996), we offer an in-depth description of how this family realized such innovative outcomes. Our description is structured around a typology of antecedent factors. In consideration of Mintzberg and Waters' (1985) distinction between the intentional and emergent dimensions of strategy, and the recurring observation (Van de Ven 1988; Kanter 1988; King and Anderson 1995) of innovation often occurring in a more emergent rather than controllable and predictable fashion, we include both intentional and unintentional antecedents. In recognition of the centrality of the intergenerational characteristic of family enterprise noted earlier, we include both elder and younger generations as significant actors. Combination of the dimension of intent with the bi-generational actor set, results in the framework represented in Figure 3.

Intended antecedent behavior by parent: "You will take piano lessons!"

A logical starting point is with the intent of the parental generation. In the case of the Brubeck family, such initiatives included mandatory piano lessons for five of the children. As son Chris remembers, "at one point there was one piano teacher who came to the house and gave five lessons in a row, each one a half-hour" (Hall 1996,103). Given the necessity of lessons being supported by practice, an additional parental initiative was also undertaken. According to Chris, "Dave finally got into the bribery system where he'd give each of us a quarter for every time we'd seriously practice for half an hour" (1996,103). However, beyond the explicit activities of lessons and practice was another, more intangible, but no less intentional, antecedent: parental encouragement.

Figure 3: Antecedents to family innovation: By generation and intent

	Elder generation	Younger generation
Intentional antecedent	Intentional behaviors by elder generation (mandatory piano lessons)	Intentional behaviors by younger generation (instrumental preference)
Unintentional antecedent	Unintentional behaviors by elder generation (serendipitous introductions)	Unintentional behaviors by younger generation (play activities)

As Darius remembers, "I didn't feel pushed, but there was a lot of affirmation that I had talent and I had to live up to that" (Hall 1996, 99).

Unintended antecedent behavior by parent: "How'd they learn that?"

The aforementioned piano lessons were not provided with the intent of launching the children's musical careers. Indeed, as Dave recounts, he wanted to impart only an aesthetic appreciation for the art form, not a career calling within it:

> I did not plan for my children to become professional musicians. By the time they were of an age to study music, I was fully aware of how difficult life as a jazz musician could be. However, I wanted them to have music as a fulfilling part of their lives, so their piano studies began early, as did mine. (Gloyd 1997, 9)

However, four of his children deliberately chose to pursue careers in music. Building on Mintzberg and Waters' (1985) observation that strategies are not always explicitly chosen, but sometimes emerge despite conscious choices to the contrary, we propose that part of the senior Brubeck's influence was unintentional. Commentary from the Brubeck's children supports this possibility, with such unintended influence occurring during the normal course of their father's professional involvements. Chris, who eventually majored on trombone as his instrument of choice, recounts one such poignant moment, when Dave unintentionally facilitated a particularly influential encounter:

> I remember Dave took me to a concert on Long Island where he was double-billed with Louis Armstrong. Dave introduced me

> to Louis and said, "This is my son Chris and he's learning to play trombone." Now I have really fat lips. Louis said, "Oh yeah? You got the chops for it!" That really felt like the pope's blessing, as far as I was concerned. (Hall 1996, 104)

Intended antecedent behavior by child: "This is what I want to do!"

Given the plethora of occupational choices available, how and why did four children embrace music not simply as an avocational hobby, but as a career calling? As the children's own testimony suggests, the decisions were neither demanded nor expected by their parents, but self-determined. For some, the decisions stemmed from an inherent interest in the instrument – in the cases of Dan and Chris, an innate interest in the drums and trombone, respectively. According to Dan, "as far back as I can remember I wanted to play drums. My Mom says I used to pull pots and pans out in the kitchen and pound on them" (Hall 1996, 105). However, the pursuit of his interest was made all the easier by easy access to appropriate instruments and exemplary role models. "Dave's group used to rehearse at the house all the time, so there was always a drum set there, and I had access to them. Of course I was always a great admirer of Joe Morello, and I watched him a lot" (Hall 1996, 105).

Unintended antecedent behavior by child: "Hey, that's neat!"

One additional set of factors warrant mention. These include the younger generation's unintentional behaviors. One of these unintentional behaviors included an unintended pattern of integrating play with the formation of competence, in this case, musical. In addition, exposure to their parent's musically intensive careers appears to have also had the effect of communicating a comparatively restricted set of career options. As Matthew notes, "I didn't think there was another way to make a living" (Hall 1996, 100).

From Improvising Parent to Improvising Family: Learning from the Brubecks

Dave, Darius, Dan, Chris, and Matthew Brubeck's collective efforts in jazz improvisation teach us much about the nature of family innovation, and by extension, innovation in family firms. They collectively exemplify Kanter's assertion that innovation is the result of "micro-processes [that are] stimulated, facilitated, and enhanced... by a set of macro-level conditions" (1988, 169). Not only does each family member possess domain competence, creativity relevant skills, and intrinsic task motivation necessary for creativity (Amabile 1988), but their family culture encourages a wide range of experimentation, the expression of creative impulses and affection, and the valuing of individual family members (Satir 1972).

Micro-processes: Creativity of individual family members

Legendary bass player Charles Mingus once observed, "[You] can't improvise on nothin' man. You gotta improvise on somethin." His maxim concurs with Amabile's (1988) componential model of creativity that proposes that individuals must possess a "set of response possibilities" (1988,130) comprised of factual knowledge, technical skills, and talents within a particular domain. Individual members of the Brubeck family clearly demonstrate such knowledge, skill, and talent, acquired through early musical experiences and subsequent training.

However, while domain-relevant skills are a necessary component of creativity, they are not sufficient. As Amabile notes, they must be matched with creativity-related skills such as cognitive styles that favor new perspectives, the application of heuristics for exploring new cognitive pathways, and an energetic working style (Amabile 1988). In observing and listening to the Brubecks, an intra-familial tendency to welcome such exploratory behaviors is readily apparent. Oldest son Darius comments:

> Even though the music is deeply familiar, almost subconsciously so for me, rehearsals take a long time because we all contribute ideas – anything is worth a try – and because Dave keeps astonishing me with the depth and breadth of his creativity. There is probably overall more humor and risk taking. In two words: it's fun" (Gloyd 1997, 5).

Intrinsic task motivation is also an essential component in creative endeavors. It essentially provides the driving force to use and apply one's competencies in the creative process. Izard's (1977) notions of interest-excitement and joy as the basis of intrinsically motivated behavior, and Csikszentmihalyi's (1975) concept of flow seem particularly prevalent in the Brubecks' collaboration. In describing his experience of working with his brothers and father, Chris notes that "when I play with my blood brothers there is an extra dimension of joy and pride when they conjure up an extraordinary musical passage" (Gloyd 1997, 7). Clearly, working with and for the family members is not just about the extrinsic motivation of receiving a paycheck, but part of a larger, more encompassing process that provides a keen sense of enjoyment and satisfaction to each individual.

Macro-level conditions: Family culture

But how, one might question, did family members come to develop such orientations to their work and be able to work so well together? This is where family culture becomes particularly important. Family culture parallels Kanter's (1988) notion of macro-level conditions that stimulate, facilitate, and enhance innovation. From their earliest ages, the Brubeck children were exposed to their father's love of music and given every opportunity to learn it for themselves. However, there appears to have been minimal coercion in favor of making life choices that included careers in music. Time and again Darius, Chris, Dan, and Matthew indicated that the choices they made vis-à-vis their formal education, preferred instruments, musical styles, and even mentors, were self-determined. Furthermore, each son was allowed, and even encouraged, to make his own mark in the world by gaining musical experience outside the family.

The psychological consequence of this type of family freedom and flexibility was to imbue each son with a solid differentiation of self (Bowen 1966). They had a rich family environment growing up, but were further encouraged to develop their own autonomous identities. The fruit of this growth process is reflected in Chris's description of his relationship with his father: "I feel we have transcended the usual father-son relationship. We now interact as two musicians whose abilities we happily respect, and who look out for each other" (Gloyd 1997, 6). Though still father and son, the two are also mature adults, respectful and appreciative of each other's competencies and the relationship they share. This is essential to normative development (and by extension intra-family innovation) because, as Bowen indicates, individuals with low self-differentiation either conform to family demands or achieve pseudo-independence through rebellion and emotional cutoff.

In addition to a family atmosphere that encouraged such freedom and self-differentiation, interviews with the father and sons indicate the presence of Satir's (1972) idea of the normal family: one that nurtures its members, encourages listening, and mutual consideration, expresses love and affection, and promotes the feeling of being valued. Listening to the brothers' comments about each others' achievements and talents, as well as the sincere admiration they hold for each other, underscores the great level of affection they share. Chris' description of his brothers' continued growth is telling:

> On these sessions, after not recording with Darius for years, since he has been teaching in Africa, I was impressed with how he has continued to grow as a pianist. Matthew, my youngest brother by ten years, keeps coming up with original expressions. Dan, with whom I have played most often, did not really surprise me. I know how great a drummer he is. His advancement is not in what he plays these days, but in the spaces he leaves open. His creativity and technique as a soloist are still inspiring and thrilling to my ears (Gloyd 1997, 7).

Instead of feelings of rivalry, insecurity, and threat or an emotional climate of deadness that often plagues troubled families (Satir 1972; Whitaker and Keith 1981), the Brubecks seem to overflow with affection and mutual valuing. They do not stay together merely out of duty or habit, but indeed, out of their shared love of music and love for each other. Such an atmosphere is likened to Kanter's (1988) emphasis on the importance of integration versus isolation as an important macro-level condition for innovation.

From Improvising Family to Innovative Family Firm: Key Insights into Family Firm Innovation

What insights can we glean from the Brubeck family's experiences to assist us in better understanding innovation in family firms? Drawing on our distinctions between elder and younger generations, and intentional and unintentional behaviors, we propose four central insights. First, the Brubeck children would not have been able to enter into the innovative experience of jazz improvisation had their parents not intentionally equipped them with the tools, in this case the musical competencies, necessary to do so. In the context of the family firm, this suggests that intergenerational and intra-organizational innovation must be preceded by deliberate initiatives that facilitate the formation of relevant competencies in the succeeding generation. However, parental effort is also fundamentally constrained. Unless the child pursues his or her individuational trajectory, the result of initiatives, such as those undertaken by the elder Brubecks will result in nothing more than static competence devoid of creative endeavor.

The experience of the Brubeck family also points toward a second insight; namely, that parental relinquishment is a critical antecedent. Dave and Iola Brubeck, paraphrasing the Arabian proverb, both loved their children and sent them on their respective travels. They affirmed each of their children as special individuals, capable of a unique calling and life contribution. Such parental orientation serves as a necessary springboard for individuation and, by extension, ideation. However, it also suggests a paradox: the emergence of intra-familial creativity may require that the parent pay the price of the child's purposeful absence for an undetermined period. Only by exploring their interests and calling can the children come to discover their place in the world.

This latter point suggests a third insight in building a theory of family firm innovation; namely, that responsibility for intergenerational creativity also resides in large part with the younger generation. To paraphrase Joel, unless "young men see visions" there can be no innovative impetus beyond the present. Just as improvisation is about making something new of something old, so family firm creativity is about the next generation enacting new customer value propositions that either leverage past generation's achievements or, in the event of past generational failure, dispense with them and reformulate, as shown by Ted Turner.

A fourth and final principle concerns the nature of success. Nowhere in the narratives studied was there any hint of Dave or Iola's disappointment in only four of their six children pursuing careers in music. The elder Brubecks never defined the two non-musicians as failures, nor defined the four musicians as successes simply because they were following in their father's footsteps. Rather, they embraced a more synoptic view that recognized the possibility of the two pursuing their life callings, and by that measure, they were no less successful than their other siblings. By extension, a reversal of this logic also means that the four pursuing musical callings could conceivably be falling short if their decision to enter had been motivated only by perceived obligation rather than heartfelt opportunity (Rogal, 1989).

Next Steps: Avenues for Further Research

The Brubeck family's story teaches us much about what family firm innovation might involve. Their experience, together with that of the family firms mentioned earlier, also suggests some possible avenues for further research on the topic of family firm innovation. In our concluding section we sketch out several topics warranting further investigation.

Definitional issues in family firm innovation

A first issue warranting further exploration concerns the definition of innovation. Just as researchers have struggled with defining what exactly constitutes a bona fide innovation, so family firm researchers will need to wrestle with defining what constitutes a legitimate family firm innovation. For example, consider the earlier cited experiences of Robert Sarnoff at RCA, as well as the more recent strategic initiatives of the Seagram Corporation under the third generation leadership of Edgar Bronfman Jr. (Heinrich 2000). In both cases, family members have pursued acquisition-intensive strategies that relied less on new and creative initiatives than on simple asset reshuffling. Should such actions be considered family firm innovations, or rather their substitute?

Dimensional issues in family firm innovation

Related to the definitional issue are dimensional issues. As Parsons (1960) noted, every organization exists at technical, managerial, and institutional levels; while the technical refers to the core internal conversion processes and the managerial relates to the effective management of resources into and out of the organization, the institutional refers to issues of underlying legitimacy (Selznick 1957). This distinction is of particular significance when considering the evolving histories of such family enterprises as those of the Rockefeller (Collier and Horowitz 1976) and Oppenheimer clans. In the case of the former, founder John D. Rockefeller achieved unimaginable success in the emerging petroleum industry; however, his success brought with it widespread resentment and perceived illegitimacy. His son, J. D. Rockefeller Jr., undertook full-time philanthropy as his career involvement; while at least part of his motivation appears to have originated in his Christian convictions, his initiatives provided institutional legitimacy lacking under his father's leadership. Likewise, while Ernest Oppenheimer took bold and often cutthroat measures in building the Anglo American Corporation, it fell to son Harry, and grandson Nicky, to complement technical and managerial competency with institutional viability, both in South Africa and the larger international arena (Pallister, Stewart, and Lepper 1987). Do such unconventional, but no less essential, initiatives also qualify as family firm innovations?

Of mothers and daughters: Female perspectives on innovation

It would be highly remiss not to point out that the cases used as examples in this paper focused on father-son relationships – particularly, when Joel also prophesied that "your sons and *daughters* will prophesy" (Joel 2:28b, New American Standard Bible 1977: emphasis added), a radical notion for ancient Hebrew civilization. As such, future research would also benefit by investigations of mother- and daughter-controlled firms. Would innovation look different in such firms and be motivated by similar dynamics? And for that matter, what about mixed father-daughter (such as between Hugh and Christie Hefner at Playboy Corporation) and mother-son (as between Katherine and Donald Graham at The Washington Post) combinations?

Intra- versus Inter-generational issues

Another important issue concerns differences in intra- and intergenerational family firm innovation. In this paper we have expressly focused on intergenerational dynamics; however, there is no shortage of family firms comprised of multiple siblings involved in their management. One need only consider the experiences of Roy and Walt Disney (Potts and Behr 1987), Albert, Paul, Edward, and Ralph Reichmann (Blanco 1997), or Wallace and Harrison McCain (Waldie and Jennison 1996). In each of these cases siblings effectively led their organizations in formulating innovative initiatives. However, as Davis (1982) observed in his study of father-son dyads, with the introduction of intergenerational dynamics also come perilous individuational interactions. Might this signal that inter-generational family innovation is sufficiently perplexing to be beyond the capabilities of most family firms?

The significance of family culture

Another more macro-level research point that warrants further investigation concerns the role of family culture. Given observed differences between families' structure (Minuchin 1974), a more comprehensive theory of family firm innovation must grapple with the influence of a family's system dynamics and innovative behavior. In particular, what kinds of family cultures support innovation and what cultures frustrate it? Furthermore, how do particular aspects of family culture, such as structure, emotional climate, and intergenerational projection processes, impact on innovation?

Moving beyond the family: The role of non-family employees

Given the significant contribution made to many family firms by non-family members, another key issue concerns the inclusion of non-family stakeholders in the innovation process. As Teal and Willigan's (1989) event study of one family's reaction to advances initiated by an extremely competent non-family manager suggest, one of the fundamental challenges that remains perpetually relevant to any family firm is reckoning with the possibility of a younger generation that possesses no, or a simply underdeveloped, vision. Hence the question: How do family firms manage to include non-family in their creative initiatives and still remain family firms?

Family firm size and innovation

While the Brubeck case study consisted of father and four sons, it would be interesting to further research the role firm size plays in the dynamics of innovation. For instance, are family culture and individual member creativity more pronounced in their effects on innovation in small and medium-sized firms compared to larger ones? If so, are these differences of mere magnitude or are they qualitatively different? This would be a welcome addition to the innovation literature, because although firm slack has been a variable of interest in explaining organizational innovation, firm size has been less often considered.

Conclusion: Fulfilling Joel's Prophecy

This article sought to advance insight into the intersection of innovation and family business. Its basic thesis is that innovative endeavor in family firms is possible, but for many reasons never realized. However, given anecdotal examples of family firms achieving innovative outcomes (Goldwasser 1986), the possibility remains. Our study of the Brubeck family hints at several helpful clues central to resolving this riddle. Specific behaviors, both intentional and unintentional, by both elder and younger generation, figure in; however, underlying all is a common shared interest in the purpose and nature of the innovative endeavor.

In *Identity and the Life Cycle*, Erikson (1980) used the term "generativity" in the seventh stage of his psychosocial model of human development labeled *generativity versus stagnation*. "Generativity is primarily the interest in establishing and guiding the next generation, although there are people who ... do not apply this drive to offspring but to other forms of altruistic concern and creativity" (1980, 103). Erikson's idea of generativity could not be more fitting for the family firm. By extending this idea of establishing the next generation to organizations, generativity deals with behaviors directed at creating beneficial change for the purpose of growing organizations, their people, products, processes, and services. Perhaps this is the underlying message in Joel's words that we can take away regarding family firm innovation: family firms may have no choice but to wrestle with issues of stagnation versus generativity if they hope to set in place the processes of innovation necessary to propel them forward into future generations.

Acknowledgments

This article received the *2000 Best Unpublished Research Paper on Family Business Award* from *The Family Firm Institute* and the *2001 USASBE Best Conceptual Paper Award* from *Entrepreneurship: Theory and Practice*. A later version of this article appears in *Family Business Review* (2001, Vol. 14, No. 4). Finally, the authors also wish to acknowledge the contributions made by students in the undergraduate course in *Family Business Management* offered during the winter term of the 1999-2000 school year at the I. H. Asper School of Business at the University of Manitoba.

References

Alcorn, P.B. (1982). *Success and Survival in the Family-Owned Business*. New York: Warner Books.

Amabile, T. (1983). *The Social Psychology of Creativity*. New York: Springer-Verlag.

Amabile, T. (1988). A model of creativity and innovation in organizations. *Research in Organizational Behavior* 10, 127-67.

Anderson, N. (1992). Work group innovation: a state-of-the-art review. In *Organizational Change and Innovation*, ed. D. Hosking and N. Anderson. London: Routledge.

Anderson, N., and King, N. (1991). Managing innovation in organizations. *Leadership and Organizational Development Journal* 12: 17-21.

Anderson, N., and King, N. (1993). Innovation in organizations. *International Review of Industrial and Organizational Psychology* 8: 1-34.

Anderson, N., and West, M. (1998). Measuring climate for work group innovation: development and validation of the team climate inventory. *Journal of Organizational Behavior* 19: 235-58.

Angle, H. (1989). Psychology and organizational innovation. In *Research on the Management of Innovation*, ed. A. Van de Ven, H. Angle, and M. Poole. New York: Harper and Row.

Bastien, D. T., and T. J. Hostager. (1988). Jazz as a process of organizational innovation. *Communication Research* 15: 582-602.

Bianco, A. (1997). *The Reichmanns: Family, Faith, Fortune and the Empire of Olympia and York.* Toronto: Random House.

Bowen, M. (1966). The use of family theory in clinical practice. *Comprehensive Psychiatry* 7: 345-74.

Chakravarthy, B. (1986). Measuring strategic performance. *Strategic Management Journal* 7: 437-58.

Chua, J., Chrisman, J., and Sharma, P. (1999). Defining family business by behavior. *Entrepreneurship Theory and Practice* 23(4): 19-38.

Collier, P., and Horowitz, D. (1976). *The Rockefellers: An American Dynasty*. New York : Holt, Rinehart & Winston.

Covin, T. J. (1994). Perceptions of family-owned firms: The impact of gender and educational level. *Journal of Small Business Management* 32(3): 29-39.

Csikszentmihalyi, M. (1975). *Beyond Boredom and Anxiety*. San Francisco: Jossey-Bass.

Damanpour, F. (1991). Organizational innovation: A meta-analysis of effects of determinants and moderators. *Academy of Management Journal* 34: 555-90.

Davis, J. (1982). *The Influence of Life Stage on Father-Son Work Relationships in Family Companies*. Ph.D. Dissertation, Harvard University.

Dess, G. G., and R. B. Robinson. (1984). Measuring organizational performance in the absence of objective measures: The case of the privately held firm and conglomerate business unit. *Strategic Management Journal* 5: 265-73.

Donnelly, R.G. (1964). The family business. *Harvard Business Review* 42: 93-105.

Downs, G., and Mohr, L. (1976). Conceptual issues in the study of innovation. *Administrative Science Quarterly* 21: 700-715.

Drazin, R., and Schoonhoven, C. (1996). Community, population, and organizational effects on innovation: a multilevel perspective. *Academy of Management Journal* 39: 1065-83.

Drucker, P. (1985). *Innovation and Entrepreneurship*. New York: Harper and Row.

Dyer, W. G., and Sánchez, M. (1998). Current State of Family Business Theory and Practice as Reflected in Family Business Review 1988-1997. *Family Business Review*. 11: 287-95.

Erikson, E. (1980). *Identity and the Life Cycle*. New York: W. W. Norton & Co.

Gardner, J. W. (1964). *Self-renewal: The Individual and the Innovative Society*. New York: Harper Colophon Books.

Glassman, E. (1986). Managing for creativity: back to the basics in R & D. *R & D Management* 16: 175-83.

Gloyd, R. (1997). Notes From the Producer. From the Telarc Recording: In Their Own Sweet Way. CD-83355.

Goldwasser, T. (1986). *Family Pride: Profiles of Five of America's Best-Run Family Businesses*. New York: Dodd, Mead and Co.

Guilford, J. (1963). Intellectual resources and their values as seen by scientists. In *Scientific Creativity*, ed. C. Tayloer and F. Barron. New York: Wiley.

Gumpert, D., and Boyd, D. (1984). The loneliness of the small-business owner. *Harvard Business Review.* 62(6): 18-20, 22, 24.

Guth, W., and Ginsburg, A. (1990). Guest Editor's introduction: Corporate entrepreneurship. *Strategic Management Journal* 11: 5-16.

Hall, F. M. (1996). *It's About Time: The Dave Brubeck Story*. Fayetteville: University of Arkansas Press.

Heinrich, S. (2000, March 28). Market bets on Seagram breakup. *Financial Post*: C1, C8.

Hoy, F., and Verser, T. (1994). Emerging business, emerging field: Entrepreneurship and the family firm. *Entrepreneurship: Theory and Practice* 19(1): 9-23.

Iacocca, L., and Novak, W. (1984). *Iacocca*. Toronto: Bantam Books.

Izard, C. (1977). *Human Emotions*. New York: Plenum.

Kanter, R.M. (1983). *The Change Masters*. New York: Touchstone.

Kanter, R.M. (1988). When a thousand flowers bloom: Structural, collective, and social conditions for innovation in organization. *Research in Organizational Behavior* 10: 169-211.

Kets de Vries, M. F. (1993). The dynamics of family-controlled firms: The good and the bad news. *Organizational Dynamics* 21(3): 59-71.

Kimberly, J. (1981). Managerial innovation. In *Handbook of Organizational Design*, ed. P. Nystrom and W. Starbuck, Vol. 1. Oxford: Oxford University Press.

King, N. (1990). Innovation at work: the research literature. In *Innovation and Creativity at Work*. ed. M. West and J. Farr. Chichester: John Wiley and Sons.

King, N. and Anderson, N. (1995). *Innovation and Change in Organizations*. London: Routledge.

Kirton, M. (1976). Adaptors and innovators: A description and measure. *Journal of Applied Psychology* 61: 622-29.

Litz, R. A. (1997). The family firm's exclusion from business school research: Explaining the void, addressing the opportunity. *Entrepreneurship: Theory and Practice* 21(3): 55-71.

Maute, M., and Locander, W. (1994). Innovation as a socio-political process: an empirical analysis of influence behavior among new product managers. *Journal of Business Research* 30: 161-74.

McQueen, R. (1998). *The Eatons*. Toronto: Stoddart Publishing Co. Limited.

Michael, R. (1979). How to find-and keep-creative people. *Research Management* September: 43-45.

Mintzberg, H. and J. Waters. (1985). Of strategies, deliberate and emergent. *Strategic Management Journal* 6: 257-72.

Minuchin S. (1974). *Families and Family Therapy*. Cambridge: Harvard University Press.

Morgan, G. (1986). *Images of Organization*. Newbury Park: Sage Publications.

Morris, M. H., Williams, R.O., Allen, J.A., and Avila, R. A. (1997). Correlates of Success in Family Business Transitions. *Journal of Business Venturing* 12: 385-401.

New American Standard Bible (1977). Anaheim: J. B. McCabe.

Pallister, D., Stewart, S., and Lepper, I. (1987). *South Africa Inc.: The Oppenheimer Empire*. London: Simon and Schuster.

Parsons, T. (1960). *Structure and Process in Modern Societies*. Glencoe, Illinois: Free Press.

Potts, M., and Behr, P. (1987). *The Leading Edge*. New York: McGraw-Hill.

Quinn, J. (1988). Innovation and corporate strategy: Managed chaos. In *Readings in the Management of Innovation*, ed. M. Tushman and W. Moore. New York: HarperBusiness.

Rogal, K. (1989). Obligation or opportunity? How can could-be heirs assess their position? *Family Business Review* 2: 237-55.

Rogers, E. (1983). *Diffusion of Innovations*. 3rd ed. New York: Free Press.

Sadler-Smith, E., and Badger, B. (1998). Cognitive style, learning and innovation. *Technology Analysis and Strategic Management* 10: 247-65.

Satir, V (1972). *Peoplemaking*. Palo Alto: Science & Behavior Books.

Schumpeter, J. (1942). *Capitalism, Socialism, and Democracy*. New York: Harper and Row.

Selznick, P. (1957). *Leadership in Administration: A Sociological Interpretation*. Berkeley: University of California Press.

Sonnenfeld, J.A. (1988). *The Hero's Farewell: What Happens When CEOs Retire*. Oxford: Oxford University Press.

Stafford, K., Duncan, K.A., Danc, S., and Winter, M. (1999). A Research model of sustainable family businesses. *Family Business Review* 12: 197-208.

Staw, B. (1990). An evolutionary approach to creativity and innovation. In *Innovation and Creativity at Work*, ed. M. West and J. Farr. Chichester: John Wiley and Sons.

Teal, T., and Willigan, G. (1989). The outstanding outsider and the fumbling family. *Harvard Business Review* 67(5): 4-11.

Thompson, A.A. (1996). Turner Broadcasting System in 1992. In *Strategic management: Concepts and cases*. 9th ed.., by A.A. Thompson, Jr. and A.J. Strickland, III Chicago, Ill.: Irwin. 321-44.

Tushman, M., and Nelson, R. (1990). Introduction: technology, organizations, and innovation. *Administrative Science Quarterly* 35: 1-8.

Van de Ven, A. (1988). Central problems in the management innovation. In Tushman, M. and Moore, W. (Eds.) *Readings in the Management of Innovation.* 2nd ed., ed. M. Tushman and W. Moore. New York: HarperBusiness.

Van de Ven, A., Angle, H., and Poole, M. (1989). *Research on the Management of Innovation.* Grand Rapids: Ballinger.

Waldie, P. and Jennison, K. (1996). *A House Divided: The Untold Story of the McCain Family.* Toronto: Penguin Books Canada.

Watson, T. J., Jr. (1990). *Father Son and Co.: My Life at IBM and Beyond.* New York: Bantam Books.

Weick, K. (1976). Educational organizations as loosely coupled systems. *Administrative Science Quarterly* 21: 1.

West, M. (1990). The social psychology of innovation in groups. In West, M. and Farr, J. (Eds.) *Innovation and Creativity at Work.* Chichester: John Wiley and Sons.

West, M., and Farr, J. (1989). Innovation at work: psychological perspectives. *Social Behavior* 4: 15-30.

Whitaker, C., and Keith, D. (1981). Symbolic-experiential family therapy. In *Handbook of Family Therapy*, ed. A Gurman and D. Kniskern. New York: Brunner/Mazel.

Wolfe, B. (1994). Organizational innovation: Review, critique and suggested research directions. *Journal of Management Studies* 31: 405-31.

Young, J. (1998). *Forbes Greatest Technology Stories.* New York: John Wiley and Sons.

Zaltman, G., Duncan, R., and Holbek, J. (1973). *Innovations and Organizations.* New York: John Wiley and Sons.

Chapter 14

Family and Non-Family Priorities in Family Firms: Preliminary Evidence

Jess H. Chua, James J. Chrisman, and Pramodita Sharma

Abstract

A large portion of the family business literature is devoted to succession and other issues related to family members. Our survey indicates that issues concerning non-family managers are as important as succession planning to family firms. Based on our review of the literature and the results obtained, we conclude that family business researchers have not paid enough attention to these issues. We also present exploratory results showing how firm characteristics may influence the different concerns. Specifically, we find the concern about issues related to non-family managers to be affected by the size of the business, the number of non-family managers in the business, and the preponderance of non-family managers among the potential successors.

Introduction

It is no secret that succession issues dominate the family business literature. Commenting on the family business literature, Wortman states that "currently the field is based largely on the succession literature, where the number of conceptual and empirical studies has doubled in the past five years" (1994,3). The primary objective of this study is to determine whether academic researchers have misplaced their emphasis. Are the issues on which researchers have spent their time and efforts and to which academic journals have devoted a large portion of space truly the issues considered most important by family firms? A second objective is to try to develop an understanding about what determines the perceived importance of these issues among family firms.

The results of our questionnaire survey of Canadian family firms show that issues related to non-family managers and succession planning are tied as the most important concerns of family firms. Maintaining control over the family's financial resources is third, issues related to exit fourth, and day-to-day management of family involvement in the business ranks last. We also present exploratory evidence showing the different family firm characteristics that may influence these concerns.

In the next section we present a brief literature survey discussing the topics that occupy family business researchers and, in more detail, the role of non-family managers. Following that, we explain the methodology used to identify the concerns underlying issues perceived to be important by family firms and the ranking of these concerns. Then, we show some exploratory results linking concerns to firm characteristics. We make our conclusions in the last section.

Issues Emphasized in the Family Business Literature

In their annotated bibliography of family business management research, Sharma, Chrisman and Chua (1996) tabulated the topics covered in 226 articles on family business management. Based on their annotations, we constructed Table 1, which shows a grouping of topics and the number of articles addressing the issues in each grouping.[1]

As the table shows, succession is the most prevalent topic in the academic literature. Out of the 226 articles reviewed, 44 (19.5%) are directly concerned with succession. The next most prevalent topic is professional advice for family firms (43 out of 226 or 19.0%). Although these articles may indirectly benefit family businesses, it is likely that they benefit consultants more. The third most numerous group of articles (19 out of 226 or 8.4%) is related to distinguishing different types of family firms (16 out of 226 or 7.1%), an issue of importance to theory and research but not likely to rank high in terms of urgency for family business managers. The fourth most common topic concerns the boards of directors (13 out of 226 or 5.8%), especially the issue of having outside directors on the board of the family firm.

The dominance of succession as a research topic is not surprising. The topic is what most mainstream academic researchers believe to be

[1] This grouping and the counts are different from that used by Sharma, Chrisman, and Chua (1996) whose purpose was to discuss the literature in the context of strategic management.

Table 1: Prevalence of Topic in Academic Literature

Topic	Number of articles	Percentage of Total
Succession	44	19.5%
Professional advice	43	19.0%
Types of family business	19	8.4%
Board of directors	13	5.8%
Transition toward professional management	10	4.4%
Life cycle stages of family business	10	4.4%
Methodological issues	10	4.4%
Ethnicity	10	4.4%
Types of families	9	4.0%
Other issues	58	25.7%
Total	**226**	**100.0%**

Based on Sharma, Chrisman, and Chua (1996, Table 1)

the only thing studied by family business researchers. Our conjecture about this phenomenon runs as follows:

Family firms are known for their penchant for privacy. They are very reluctant to go outside the immediate circle of family members for advice. A trusted accountant, lawyer, or banker may, by necessity, be allowed to know the true financial picture of the family and develop a long-term relationship with the family. Otherwise, family issues are for the family only. At the time of succession, however, the incumbent confronts a problem with which he or she has little to no experience. Although many incumbents avoid facing the succession problem (Handler and Kram 1988; Leon-Guerrero, McCann, and Haley 1998), some realize that confronting the issue is important to the long-term health of the enterprise. In many cases, however, incumbents or other family members recognize their own limitations in dealing with succession and look for outside help, either from an academic or a professional consultant. This is often the only time when academics and professional consultants come into contact with family firms and have access to their data and information. As a result, the conceptual lens on family firm problems is limited to succession issues. Not surprisingly, this has led to the development of a large number of ideas, insights, frameworks, hypothesis, and recommendations about succession.

Our conjecture, if true, suggests that succession may be a dominant theme in the academic literature due to convenience, coincidence, and data access, not necessarily because of its importance to the family firm.

Whether this is true has not been tested. As discussed above, the purpose of this paper is to report on a survey of the issues considered important by family firm managers and how the results compare with the issues that are emphasized in the family business literature.

Non-family manager issues are often discussed within the context of moving the family firm toward professional management. Even then, the focus is frequently on how the incumbent family member CEO should adapt to working with non-family managers. For example, Hofer and Charan (1984), Perrigo (1975), Mathews (1984), Goffe and Scasse (1985), and Firnstahl (1986) all address the importance of delegation by the CEO if professionalization of management is to succeed. These authors also discuss how difficult it is for family firms to delegate to non-family managers. Dyer (1989) points out that non-family managers may also need help in adapting to professional management. In a minor observation in their paper, Post and Robins (1993) suggest that non-family members may obstruct succession in order to delay or avoid change.

Recently, there has been more attention paid specifically to non-family managers. Poza, Alfred, and Maheshawi (1997) present evidence showing that non-family managers are less satisfied with the equity of compensation and with how their firms handle growth, especially in terms of setting up systems and practices to facilitate growth. They believe more strongly than family members that the successor will be a family member and that they are excluded from senior positions. Poza, Alfred, and Maheshawi also suggest that because family firms tend to manage in a way that maximizes flexibility and avoids redundancies, non-family managers must be multi-skilled. They also show, however, that non-family members are paid a premium of 15.4% to 29.5% to compensate for these differences.

Ward (1997) presents his opinions about how to attract and retain excellent non-family managers. His list of initiatives includes emphasizing merit, giving non-family managers opportunities to accumulate personal wealth, and assuring career opportunities for the best among them. Gallo and Vilaseca (1998) provide data showing that firm performance is not affected by whether or not the CFO is a family member.

While these recent articles have started to focus some attention on the role of the non-family managers in family firms, there is no evidence to show how important family firms perceive the issues related to non-family managers to be. We are also unaware of any research results showing what affects the perceived importance of such issues to the family firm. The purpose of this study is to address these gaps in the literature.

Survey on Family Firm Issues

Empirical research on family business has had to rely on convenience samples such as the membership lists of professional associations or the mailing lists of family business consultants, and this study is no exception. For our survey, we use the mailing list of the Canadian Association of Family Enterprises (CAFE) and a random sample of 500 family firms from the more than 4,000 names on the mailing list of Deloitte and Touche (Canada). CAFE is the largest national association of Canadian family firms, with a membership of 585 (approximately 10% are family business consultants and bankers) at the time of data collection (Fall, 1994). Included in its mailing list were approximately 1,000 non-member family firms.

Chua, Chrisman, and Sharma (1999) present a theoretical definition of the family firm that includes two critical characteristics. The first is that the firm's desired future must be developed by the family; the second is that there must be the potential for this desired future to be pursued across generations of the family. Realizing the difficulty in using the theoretical definition as a criterion when performing empirical research, they examine the relationship between their definition and the various combinations of family involvement in the family firm. The one combination of family involvement that makes the business unequivocally a family business, under their definition, occurs when the business is both owned and managed by the family.

After eliminating duplications and names that were obviously not family firms (e.g., account managers in banks, lawyers in large law firms, etc.), questionnaires were sent to 1,725 family firms. This included 483 CAFE members and 1,242 non-CAFE firms. Two mailings of the questionnaire were sent to the sample of firms; 211 questionnaires from CAFE members and 274 from non-CAFE members were returned. This yielded a response rate of 44% for CAFE members, 22% for non-members, and 28% overall. Nearly 90% of these respondents were family members who held positions as either the president or chairman of the board of their firms. The remainder were family members who held positions as vice-president or director.

To decrease the possibility that respondents were in fact not family firms, respondents were asked to return the questionnaire if they did not consider themselves a family business. Still, this would not have ensured that the respondents qualify as family firms according to the theoretical definition of Chua, Chrisman, and Sharma (1999). Therefore, we further screened the sample and retained only those firms that submitted data indicating that they were both family-owned and family-managed. This left us with a sample of 467 firms.

The survey covered several topics of research. The questions and sections relevant to this study are presented in the Appendix. As shown in the Appendix, we asked respondents to rate a number of issues on a seven-point Likert scale. The list of issues was based on our interpretation of the topics covered in the academic literature. We did not include some research, for example, types of family and ethnicity, in the survey because, although we concede that these topics are important, we could not imagine how they would be important *management* issues for a family firm. Also collected were firm-specific data pertaining to size, age, number of employees, potential successors, and participation of non-family managers in the business.

Importance Ranking of Issues

Our tests for bias found none according to CAFE membership and timing of response, i.e., there were no significant differences in the responses of CAFE members and non-members, nor were there significant differences between the responses received from the first and second mailings of the questionnaire. Almost all of the firms used in the analysis were 100% owned by the family. Not one was less than 90% owned by the family. All of the firms were family-managed since this was one of the screening criteria. Therefore, we concluded that all of these firms qualified as family firms and, likewise, that all of our respondents were family members holding high ranking managerial positions in the business.

The CAFE firms were mainly managed by the second or third generation family members while the non-member firms were primarily managed by first generation family members. The median number of employees was 45; therefore, although not large in the global context, they were not the typical small businesses that people have in mind when they hear the term "family business."

Univariate Ranking of Mean Ratings

A ranking of the ratings for the different issues is presented in Table 2. The table also shows the standard deviation as an indication of the agreement among respondents. For all variables, the minimum rating was 1 and the maximum 7.

The five most highly rated issues are: STLT (Balancing short-term and long-term business decisions), ESTAX (Planning for estate taxes),

Table 2: Ranking of Average Ratings for Issues

Rank	Variable	Mean	Std. Dev.	Mean is significantly higher* than those for variables
1	STLT	5.17	1.41	3 - 26
2	ESTAX	5.00	1.78	5 - 26
3	BAL	4.91	1.54	7 - 26
4	PRPSUC	4.88	1.94	7 - 26
5	CNTRL	4.81	1.97	9 - 26
6	LOYAL	4.75	1.93	9 - 26
7	SELSUC	4.68	2.08	9 - 26
8	INVLV	4.63	1.89	9 - 26
9	ASTBUS	4.43	1.74	11 - 26
10	CMPFAM	4.33	1.82	12 - 26
11	SUCFAM	4.10	2.09	14 - 26
12	OWNSHP	4.07	2.04	17 - 26
13	VALUE	4.06	1.91	17 - 26
14	BODIR	3.87	1.99	18 - 26
15	SLFAM	3.86	1.87	19 - 26
16	CHGPRO	3.85	1.92	19 - 26
17	CNFLCT	3.77	2.11	20 - 26
88	DMCRAT	3.65	1.94	20 - 26
19	RLFND	3.54	2.02	20 - 26
20	INCL	3.29	1.87	24 - 26
21	FMFND	3.28	1.88	24 - 26
22	ASTFAM	3.25	1.88	24 - 26
23	INCOMP	3.13	2.17	26
24	RIVAL	3.01	1.96	26
25	OUTBY	2.93	1.96	26
26	BUYOUT	2.59	1.93	

* 5% level

BAL (Balancing family concerns and business interests), PRPSUC (Preparing and training a successor), and CNTRL (Maintaining ownership control in the family). STLT's importance is reinforced by its having the lowest standard deviation, indicating that there is the least disagreement about the importance of this issue. BAL has the second lowest standard deviation, while ESTAX has the fourth lowest. STLT is not highly correlated with any of the other variables. Its highest correlations

(0.39 and 0.31, respectively) are with ASTBUS (Seeking outside assistance to deal with business problems) and VALUE (Valuing the business). These univariate results suggest that the most important concern is managing the firm's strategic and financial performances in the best interests of the family.

Three of these top five rated issues (ESTAX, PRPSUC, CNTRL) are typically classified as succession issues. This suggests that the academic literature's preoccupation with succession issues may be justified. The two issues related to professional advice, ASTBUS (seeking assistance from outsiders to resolve business problems) and ASTFAM (22nd) (seeking assistance from outsiders advisors to resolve family problems), the second most frequent topic covered in the academic literature, were ranked 9th and 22nd, respectively. Both are rated lower than LOYAL (6th) (maintaining loyalty of non-family managers) and INVLV (8th) (involving non-family managers in making strategic decisions), two variables dealing with non-family managers.

Board of directors (BODIR), the fourth most discussed topic in the academic literature, is ranked 14th. Its mean is 3.87, below the midpoint of 4.0. Although the deviation from the midpoint is not statistically significant ($t = 1.4$), this result does suggest that having outside members on the board of directors may be an issue of greater importance to academics and consultants than to family firms.[2]

Concerns Underlying the Issues

To identify the broader concerns underlying the issues, we relied on factor analysis. There is no a priori reason to specify orthogonal factors. In fact, it is highly likely that the underlying concerns are correlated. For example, if the firm is experiencing few problems dealing with the family members' involvement in the business, the incumbent may think more about succession than about exit. If there are more loyal and competent non-family managers, the incumbent may be more willing to think about transition to professional management. Therefore, we decided to use an oblique (promax) rotation to identify the underlying concerns.[3]

The results of a five-factor solution are presented in Table 3. Specifying six or seven factors neither provided additional insight nor improved

[2] This may indicate that family businesses do not fully recognize the value of the board of directors rather than that the board of directors is not important.

[3] Orthogonal (varimax) rotation does not change the results qualitatively.

Table 3: Extract from Pattern Matrix for Principal Axis Extraction with Promax Rotation

	FACTOR LOADINGS				
	1	2	3	4	5
ASTBUS					0.45
ASTFAM	0.68				
BAL					
BODIR	0.46				
BUYOUT	0.54				
CHGPRO					0.47
CMPFAM	0.43				
CNFLCT	0.93				
CNTRL			0.45		
DMCRAT	0.46				
ESTAX			0.78		
INCL					
INCOMP	0.69				
INVLV		0.70			
LOYAL		0.98			
FMFND			0.51		
OUTBY					0.58
OWNSHP			0.43		
PRPSUC				0.87	
RIVAL	0.77				
RLFND					
SLFAM				0.54	
STLT					
SELSUC				0.81	
SUCFAM		0.65			
VALUE			0.50		
Variance explained	**30.7**	**8.6**	**7.5**	**5.0**	**4.5**

interpretation. Table 3 shows the pattern matrix after promax rotation and the percentage variance explained. To facilitate this discussion, only the contributions to variances above 0.40 are shown. As the table shows, the first factor explains 30.7% of the variance, the second 8.6%, the third 7.5%, the fourth 5.0%, and the fifth 4.5%, for a total of 56.3%. The Chi-square test for goodness-of-fit is highly significant.

The first factor contributes the most to the variances of CNFLCT (resolving conflicts among family members), RIVAL (dealing with rivalry among potential family member successors), INCOMP (dealing with incompetent family members in the business), and ASTFAM. These are all issues related to the day-to-day involvement of family members in the business.

The second factor contributes the most to the variances of LOYAL, INVLV, and SUCFAM (developing relationship between successor and non-family managers). These are issues concerning non-family managers.

The third factor contributes significantly to the variances of ESTAX (planning for estate taxes), FMFND (setting up a family foundation), VALUE (valuing the business), and CNTRL (maintaining ownership control within the family). These are all issues concerned with keeping control over the financial resources accumulated by the family.

The fourth factor contributes to the variances of SELSUC (selecting the successor), PRPSUC (preparing and training a successor), and SLFAM (selecting family members for positions in the business). These are components of succession planning.

The last factor contributes mainly to the variances of OUTBY (finding an outside buyer for the business) and CHGPRO (changing from family to professional management). These are issues related to the family's exit from ownership and/or management.

In summary, factor analysis suggests that there are five concerns underlying the issues considered important by the respondents to our survey. These are:

- day-to-day problems of managing family involvement in the business;
- keeping the non-family managers involved and loyal;
- maintaining control over the family's financial resources;
- selecting, preparing, and training a successor;
- issues related to the family's exit from the firm.

Importance Ranking of Underlying Concerns

Aside from identifying the concerns and underlying issues considered important by family firms, it would be beneficial to know how these underlying concerns rank in importance. Statistically, one cannot directly equate the variance explained by a factor to its importance as a concern; the ability of a factor to explain a larger portion of the aggregate variance in the data arises from a wider diversity of opinions about the importance of the issues subsumed in the factor. In other words, if the aggregated variance in the data is caused mainly by disagreements among respondents about the importance of certain issues, then the factor that includes these issues would explain the largest portion of the aggregated variance. This by no means implies that family firm CEOs consider the included issues to have the highest ranking in importance; instead, it just means that family firm CEOs differ more widely in terms of their importance rankings of these issues. To illustrate the point further, take the example of STLT as a variable. STLT has the highest mean importance rating and the lowest variance. Its low variance cannot be the cause of the aggregated variance in the data. It is also not highly correlated with any of the other issues and cannot help to explain the variances in the other variables. As a result, it is not subsumed or included in any of the five factors identified as underlying concerns. Nevertheless, its highest importance rating implies that this issue is of utmost importance to family firm managers.

To get an indication of how important each of the underlying concerns is to family firms, we constructed a set of latent ratings for each underlying concern or factor. These latent ratings provided an overall score of the underlying concerns that the factors appeared to tap. Unfortunately, it was not possible to calculate latent rating from the typical statistical package because factor analysis is performed with standardized variables to make the mean equal to zero. To obtain a latent rating for a particular factor, we calculated the average of the raw ratings for the issues with loadings on a particular factor of 0.4 or higher. For example, the latent rating for factor 2 (concern about non-family manager issues) was calculated as the average ratings of INVLV, LOYAL, and SUCFAM. In other words, for each firm, we calculated the average of that firm's ratings for INVLV, LOYAL, and SUCFAM (e.g., [5+3+6]/3 = 4.67 latent rating) to represent the importance of non-family manager issues to that firm. We used a contribution of 0.4 as the cut-off because, as shown in Table 3, this provided a non-overlapping allocation of the issues to the factors. For each firm, we calculated five latent

Table 4: Mean Latent Ratings for Underlying Concerns

Factor	Underlying Concern	Mean	Std. Dev.	Mean is significantly higher than that of factor
2	Non-family managers	4.44	1.67	3, 5, 1
4	Succession planning	4.41	1.68	3, 5, 1
3	Control over family financial resources	4.20	1.31	5, 1
5	Exit from management/ownership	3.73	1.42	1
1	Day-to-day family involvement issues	3.42	1.37	

ratings, one for each of the factors. We then averaged the latent ratings for each factor across the firms to obtain the mean latent rating for that factor and compared the mean latent rating with those for the other factors.

The results are presented in Table 4. The mean latent rating for factor 2 (non-family manager issues) is highest, although it is not significantly higher than that for factor 4 (succession planning). Both are significantly higher than those of the three other factors. The mean latent rating for factor 3 (maintaining control over financial resources) is significantly higher than those for factors 5 (exit) and 1 (day-to-day management of family involvement). The mean latent rating for factor 1 is significantly lower than those of the rest.

As the numbers suggest, one of the two most highly rated concerns is related to keeping the non-family managers involved and loyal. There is definitely a gap in our understanding of the role played by these stakeholders. What are their stakes in the business? What are their legitimate (legal as well as moral) claims on the firm after establishing their credibility and longevity? How do they contribute to the competitive advantage of family firms? Would empowering them enhance or detract from the family firm's value?

Succession planning is the other most highly rated concern, while maintaining control over the family's financial resources is ranked third. If verified by further research, this finding suggests that the goals of succession planning transcend maintaining family control over financial resources. To verify this, researchers need to identify the goals pursued by family firms when they opt for succession by a family member.

The fact that succession planning ranks so high in perceived importance, despite the opinions of many researchers and consultants that family firms tend to leave succession planning to chance, supports the observations made by authors such as Handler and Kram (1988) that there may be very strong psychological and institutional forces at work to resist succession planning efforts. On the other hand, this contradiction may also support our earlier contention that, despite recognizing the importance of succession, family firms simply do not know where to begin.

Concern about exit from management and/or ownership is ranked fourth, and day-to-day management of family involvement in the firm is last. The low ranking and latent rating (below the scale midpoint of 4.0) of the exit factor, combined with the rankings and latent ratings of the other underlying concerns, suggest that family firms are most concerned with maintaining a legacy for future generations for both economic and emotional reasons. However, the lesser importance attached to day-to-day management of family involvement suggests that our respondents value a viable business that operates for the good of the family more highly than the short-term contentment of individual family members.

In conclusion, the survey results justify the academic literature's preoccupation with succession in terms of both succession planning and maintaining control over the family's financial resources. However, the results point out that issues related to keeping the non-family managers involved, loyal, and motivated have not received the attention that they should, considering the great importance attached by family businesses to these issues.

It should be noted, however, that our study captures the perceived importance of issues at a single point in time for family firms that may be somewhat older, larger, and with more resources than the typical family firm. For example, larger firms may be more dependent on non-family managers than smaller firms. Thus, our results may not be representative of the entire population of family firms in the Canadian economy.

Exploratory Results on Determinants of Concerns

In an attempt to further understand the concerns discussed above, we tested whether they are related to firm-specific characteristics. To do this, we tested how the concerns of family businesses, as represented by the five factors, are affected by demographic characteristics of the firms. Table 5 displays descriptive statistics for the firm characteristics included in the analysis. We used ANOVAs rather than a MANOVA test because the five factors are highly correlated. The ANOVA results are presented in Table 6. The statistically significant relationships are all

Table 5: Descriptive Statistics for Firm Characteristics

Variable	N	Min.	Max.	Mean	Std. Dev.	Skewness*
AGE	432	1	174	42.25	27.79	1.39
GRSREV	462	1	8	5.60	1.60	(0.68)
FTE	455	0	4200	104.80	291.17	9.28
FAMNUM	462	0	12	3.07	1.67	1.50
NONFAM	447	0	446	6.56	24.80	13.95
RETIRE	461	0	1	0.62	0.49	(0.51)
GENRTN	464	1	6	1.87	0.92	1.04
BRDFM	415	0	1	0.85	0.26	(1.47)
PCTMLSC	411	0	1	0.64	0.40	(0.60)
PCTNNFSC	413	0	1	0.22	0.34	1.30

* All skewness statistics are significant at the 99% level

AGE	=	age of the business;
GRSREV	=	gross revenue;
FTE	=	full-time employees;
FAMNUM	=	family members active in the business;
NONFAM	=	number of non-family managers;
RETIRE	=	CEO expected to retire within 10 years;
GENRTN	=	generation of family members managing the business;
BRDFAM	=	number of family members on the board of directors;
PCTMLSUC	=	percentage of potential family successors who are male; and
PCTNNFSC	=	percentage of all potential successors who are not family members.

positive, and each shows that, on a univariate basis, the specific concern increases with the particular firm characteristic.

Starting with DTDFAM (Factor 1: day-to-day management of family involvement), we find that these concerns are influenced by GRSREV (revenues of the business), FAMNUM (the number of family members active in the firm), RETIRE (imminence of CEO's retirement), BRDFAM (the number of family members on the board), PCTMLSUC (percentage of potential family member successors who are male), and PCTNNFSC (percentage of all potential successors who are not family members). Most of these findings conformed to our expectations. For example, one's intuition would be that the larger the business, the more day-to-day family involvement problems will be perceived as important because of the larger stakes involved. Likewise, the higher the number of family members actively involved, especially if they serve on the board and

Table 6: Results of Univariate ANOVAs

	Underlying Concern				
	DTDFAM	NFMMGR	FINRES	SUCPLN	EXIT
AGE					
GRSREV	***	***	***	*	*
FTE		***	**		***
FAMNUM	***		***	***	
NONFAM		***	***		*
RETIRE	***		***	***	
GENRTN					
BRDFAM	***				
PCTMLSUC	***		***	***	
PCTNNFSC	***	***	***	***	**

All significant relationships reported are positive

*** $p < .01$
** $p < .05$
* $p < .10$

Dependent variables:

DTDFAM	=	day-to-day management of family involvement;
NFMMGR	=	concerns about involvement and loyalty of non-family managers;
FINRES	=	maintaining control over the family's financial resources;
SUCPLN	=	succession planning;
EXIT	=	concerns about exit from ownership and/or management;

Independent variables:

AGE	=	age of the business;
GRSREV	=	gross revenue;
FTE	=	full-time employees;
FAMNUM	=	family members active in the business;
NONFAM	=	number of non-family managers;
RETIRE	=	CEO expected to retire within 10 years;
GENRTN	=	generation of family members managing the business;
BRDFAM	=	number of family members on the board of directors;
PCTMLSUC	=	percentage of potential family successors who are male; and
PCTNNFSC	=	percentage of all potential successors who are not family members.

have to deal with the difficult trade-offs in making policy and deciding strategy, the greater the opportunities for conflict among family members. The closer the CEO gets to retirement, the more anxious he or she will be to have the issues resolved.

That the gender of the potential family member successors can cause more concern in the day-to-day management of family involvement is, however, a surprise. Although without further investigation we can offer no definitive explanation, this finding suggests that the competition among male family member who are potential successors exhibits itself more aggressively in the day-to-day operation of the business than is the case when potential successors are predominantly female. This is a confirmation of the suggestion by Chua, Chrisman, and Sharma (1999) that Canadian family firms may not treat their male and female potential successors the same. If aggression is, indeed, found by further research to be correlated with male potential successors, then behavior rather than, or in addition to, gender bias may explain this differential treatment.

The highly significant effect of non-family potential successors also requires some reflection to explain. Our conjecture is that the more non-family managers are considered to be potential successors, the more potential family member successors have to compete with them, resulting in disputes and dissatisfactions that necessitate the CEO's intervention.

Concerns about the loyalty and involvement of non-family managers (Factor 2) are affected by GRSREV, FTE (number of full-time employees), NONFAM (number of non-family managers), and PCTNNFSC (the percentage of all potential successors who are not family members). These are all indications of the family firm's dependence on non-family managers. The larger the business and the higher the number of non-family managers, the greater the firm's dependence on non-family managers. When this dependence extends to the possibility that a non-family manager may have to take over the CEO's position, then the dependence is critical.

The results indicate that concerns about maintaining control over the family's financial resources (Factor 3) increase with the size of those resources (GRSREV, FTE, NONFAM). They also suggest that the concerns are heightened by the number of family members involved in the business (FAMNUM). Clearly, the more imminent the CEO's retirement, the more urgent are the concerns. The relationship between these concerns and the percentage of all potential successors who are not family members (PCTNNFSC) supports the conventional agency cost argument about the segregation of ownership and management and the need for monitoring (e.g., Jensen and Meckling 1976). But why should male potential family member successors (PCTMLSUC) cause more concerns in this regard? Our conjecture is that, as Levinson (1971) observed, sons may have a greater need to prove their independence than daughters. As a result, sons may be more likely than daughters to want to change the way the business is operated (Babicky 1987), a possibility that could increase the incumbent's concerns about the family's financial resources. This result again suggests that the differences in *behaviors* of male and female family members may contribute as much to the way they are treated as does their gender. Further research is needed on this issue.

The data appear to suggest that imminence of retirement (RETIRE) makes concerns about succession planning (Factor 4) more urgent. Size (GRSREV) provides more to be handed down to the next generation. We interpret the other significant influences to mean that the more family members and potential successors the family firm has (FAMNUM, PCTNNFSC), the more it will have to plan. But, again, having more male than female potential family member successors (PCTMLSUC) is a source of the concern. These relationships support a conjecture that, in succession planning, family firms perceive that as competition among family members, especially males, and non-family members increases, it will be more difficult to implement the process successfully.

Finally, it appears that family firms think more about exit if there is more value (GRSREV) and the value is more dependent on non-family managers (NONFAM) and non-family potential successors (PCTNNFSC).

Conclusions

The objective of this study was to ascertain whether academic researchers have misplaced their emphasis and spent their time and efforts on issues considered unimportant by family businesses. We presented the results of a literature survey showing the prevalence of different topics in the academic literature. We have also presented the results of a questionnaire survey showing the issues considered most important to the incumbent leader of Canadian family firms.

Our univariate results show that balancing short- and long-term business decisions and planning for estate taxes are the most important individual issues facing the family business managers in our sample. The first finding underscores the call of Sharma, Chrisman, and Chua (1996) for a strategic management approach to family business studies. The second indicates that the fields of accounting and finance have potentially much to offer to family businesses.

These findings are enhanced by the results obtained through factor analysis and ANOVA tests. Thus, estate tax issues really just highlight a concern about keeping control of the financial resources of the family. Maintaining ownership control and establishing the value of the business are also part of this concern. The factor analysis results and ANOVA tests also make it clearer what short-term and long-term decisions most occupy the minds of our respondents. Thus, concerns with maintaining the loyalty and commitment of non-family managers and concern about succession planning are perceived to be the two most important issues. In the short term, the respondents seem to think their non-family managers are key to the success of their businesses. In the long term, providing for the firm's future leadership is of paramount importance.

The high average importance rating for succession planning and maintaining control over the family's financial resources justify the preoccupation of family business researchers with succession issues. A surprising observation, that must be corroborated by further research, suggests that concerns with succession planning may be as much a function of anticipated problems in implementation as with the need to ensure continuity of leadership or protection of the family's financial resources.

The ANOVA results show that imminence of the CEO's retirement, firm size, the proportion of non-family potential successors, and a host of other firm characteristics affect the importance of the underlying concerns. Therefore, the conclusions drawn here about the relative importance of the issues studied are contingent on the situation of the firm. For example, firm size appears to influence the significance of all the underlying concerns. The sample of firms used in this study comes from the mailing list of an association of family firms. If the size distribution of these firms is significantly different from most family businesses, then the conclusions made here may only apply to a subset of all family businesses. Future research must keep in mind that family businesses are not a homogeneous group. An important implication of this study, then, is the need to identify and classify different populations of family firms. To our knowledge, this classification problem has not been addressed in a rigorous way in any study.

Finally, the great concern for non-family managers increases with the firm's dependence on them as resource dependence and agency theories would suggest (e.g., Jensen and Meckling 1976; Pfeffer and Salancik 1978). Issues related to non-family managers have received very little

Appendix

Relevant Portion Extracted from Survey Instrument

Information About Your Business

Please answer the following questions about your business by checking the response that most closely describes your business or by specifying the information requested.

Year first sale made: ____

Gross Revenues in 1993:	_____ < $100,000	_____ $1,000,000 to $4,999,999
	_____ $100,000 to $249,999	_____ $5,000,000 to $9,999,999
	_____ $250,000 to $499,999	_____ $10,000,000 to $50,000,000
	_____ $500,000 to $999,999	_____ > $50,000,000

Number of full time employees in 1993: ________________________________

Number of family members (including you) active in the business: ________________

Number of non-family managers in the business: ________________________

Is the current president likely to retire in the next ten years? ________ Yes ________ No

Which generation of the family is operating the business today?

_____ 1st _____ 2nd _____ 3rd _____ 4th _____ 5th _____ 6th _____ 7th _____ 8th

Membership of the board of directors: (Please specify number)

Family members: __________________ Non-family members __________________

Number of persons who have the potential to assume presidency in the next ten years:

Family members:	__________ Male	__________ Female
Non-family members:	__________ Male	__________ Female

Issues for Family Business

Listed below are several issues found to be of concern to family enterprises. Please circle the response that most closely captures the importance of each of the issues to your business.

Variable	Importance of the Issue to your Company	Not important						Critically important
ASTBUS	Seeking assistance from outsiders to resolve business problems	1	2	3	4	5	6	7
ASTFAM	Seeking assistance from outsiders to resolve family problems	1	2	3	4	5	6	7
BAL	Balancing family concerns and business interests	1	2	3	4	5	6	7
BODIR	Defining the role of the board of directors	1	2	3	4	5	6	7
BUYOUT	Buying out family members not actively involved in the business	1	2	3	4	5	6	7
CHGPRO	Changing from family management to professional management	1	2	3	4	5	6	7
CMPFAM	Compensating family members involved in the business	1	2	3	4	5	6	7
CNFLCT	Resolving conflicts among family members	1	2	3	4	5	6	7
CNTRL	Maintaining ownership control in the family	1	2	3	4	5	6	7
DMCRAT	Changing from an autocratic to a democratic style of leadership	1	2	3	4	5	6	7
ESTAX	Planning for estate taxes	1	2	3	4	5	6	7
FMFND	Setting up a family foundation	1	2	3	4	5	6	7
INCL	Including non-family members in the board of directors	1	2	3	4	5	6	7
INCOMP	Dealing with incompetent family members active in the business	1	2	3	4	5	6	7
INVLV	Involving non-family managers in making strategic decisions	1	2	3	4	5	6	7
LOYAL	Maintaining loyalty of non-family managers	1	2	3	4	5	6	7
OUTBY	Finding an outside buyer for the business	1	2	3	4	5	6	7
OWNSHP	Distributing ownership among family members	1	2	3	4	5	6	7
PRPSUC	Preparing and training a successor	1	2	3	4	5	6	7
RIVAL	Dealing with rivalry among potential family member successors	1	2	3	4	5	6	7
RLFND	Maintaining a role for the founder in the business after retirement	1	2	3	4	5	6	7
SELSUC	Selecting the successor	1	2	3	4	5	6	7
SLFAM	Selecting family members for positions in the business	1	2	3	4	5	6	7
STLT	Balancing short-term and long-term business decisions	1	2	3	4	5	6	7
SUCFAM	Developing relationship between successor & non-family managers	1	2	3	4	5	6	7
VALUE	Valuing the business	1	2	3	4	5	6	7

attention by researchers, yet such concerns rank as high as succession in the minds of family business managers. We see this as an entirely new avenue of research that could enhance our understanding of family firms and improve their long-term performance.

References

Babicky, J. (1987). Consulting to the family business. *Journal of Management Consulting* 3(4): 25-32.

Chua, J., Chrisman, J., and Sharma, P. (1999) Defining the Family Firm by Behavior. *Entrepreneurship: Theory and Practice* 23(4): 19-39.

Deloitte & Touche study. (1999). $1.3 trillion at stake as Canada's family businesses face leadership crisis http://www.deloitte.ca/whatsnew/Headlines/FamilyBusiness.htm.

Dyer, W.G., Jr. (1989). Integrating professional management into a family owned business. *Family Business Review* 2(3): 221-235.

Firnstahl, T.W. (1986). Letting go. *Harvard Business Review* 64(5):14-18.

Gallo, M.A., and Vilaseca, A. A (1998) Financial perspective on structure, conduct, and performance in the family firm: An empirical study. *Family Business Review* 11: 35-48.

Goffe, R., and Scasse, R. (1985). Proprietorial control in family firms: Some functions of 'quasi-organic' management systems. *Journal of Management Studies* 22(1): 53-68.

Handler, W.C. and Kram, K.E. (1988). Succession in family firms: The problem of resistance. *Family Business Review* 1(4): 361-81.

Hofer, C.W. and Charan, R. (1984). The transition to professional management: Impossible? In *Family Business Sourcebook*, ed. C.E. Aronoff and J.L. Ward. Detroit: Omnigraphics.

Jensen, M.C., and Meckling, W.H. (1976). Theory of the firm: Managerial behavior, agency costs, and ownership structure. *Journal of Financial Economics* 3: 305-60.

Leon-Guerrero, A.Y., McCann III, J.E., and Haley Jr., J.D. (1998). A study of practice utilization in family businesses. *Family Business Review* 11:107-20.

Levinson, H. (1971). Conflicts that plague family businesses. *Harvard Business Review* 49(2): 90-98.

Mathews, G.H. (1984). Run your business or build an organization? *Harvard Business Review* 62(2): 34-44.

Perrigo, A.E.B. (1975). Delegation and succession in the small firm. *Personnel Management* May: 35-37.

Pfeffer, J., and Salancik, G. (1978). *The external control of organizations: A resource dependence perspective.* New York: Harper & Row.

Post, J.M. and Robins, R.S. (1993). The captive king and his captive court: The psychopolitical dynamics of the disabled leader and his inner circle. *Family Business Review* 6: 203-21.

Poza, E.F., Alfred, T., and Maheshawi, A. (1997). Stakeholder perceptions of culture and management practices in family and family firms: A preliminary report. *Family Business Review* 10: 135-55.

Sharma, P., Chrisman, J.J., and Chua, J.H. (1996). *A Review and Annotated Bibliography of Family Business Studies.* Boston: Kluwer Academic Publishers.

Tabachnick, B.G., and Fidell, L.S. (1989). *Using Multivariate Statistics.* New York: Harper & Row.

Ward, J.L. (1997). Growing the family business: Special challenges and best practices. *Family Business Review* 10: 323-337.

Wortman, M. S., Jr. (1994). Theoretical foundations for family-owned business: A conceptual and research based paradigm. *Family Business Review* 7: 3-27.

Chapter 15

Executive Pay Practices in Family-Controlled Firms in Canada: A Preliminary Study

Yun W. Park

Abstract

In this paper I report the executive pay practices in family-controlled firms in Canada from 1993 to 1997. I find that family-controlled firms place less weight on stock-based incentives than do widely held firms and that the total pay of executives in family-controlled firms is lower than in widely held firms. These preliminary empirical findings are discussed in light of the agency literature and the private incentives of controlling families.

Introduction

This study investigates the executive pay practices of family-controlled firms in Canada. The Canadian economy offers an ideal setting in which to examine this issue, since Canada has a large number of publicly traded firms controlled by families. As well, high-quality executive pay data is publicly available due to the comprehensive executive pay disclosure required by the Ontario Securities Commission.

Concentrated ownership is a prominent organizational characteristic in Canada, and the economic effect of concentrated ownership has been the subject of investigation in a number of Canadian corporate governance studies (Eckbo and Verma 1994; Daniels and Halpern 1996; Morck 1996; Park, Nelson, and Torabzadeh 2000; and Park, Nelson, and Huson 2001). In particular, family control has been an essential aspect of ownership concentration in Canada (Morck, Strangeland, and Yeung 1998). In *Titans*, an engaging account of the fall of old business dynasties and the rise of new ones, Peter Newman (1998) notes that, at one point in the 1950s, nine families controlled nearly half of the total shares traded on the Toronto Stock Exchange. In *Controlling Interest*, a pioneering journalistic investigation of Canadian family fortunes,

Diane Francis (1986) reports that some thirty-two families and five conglomerates controlled one-third of the nation's corporate assets as late as 1986.

When a founder-entrepreneur takes his or her firm public, he or she often assumes the role of CEO, chairperson, and controlling shareholder of a now publicly traded firm. The firm's founder eventually steps to the side and lets a successor, either from the family or from the outside, assume the active management of the firm. The controlling ownership interest often stays within the family. This process may go on for a number of generations, as in the case of the Molsons of Molson Inc. and the Thomsons of Thomson Corp. Through all of this, the family-controlled firm still needs to motivate the performance of their top executives as do firms with other governance arrangements. This paper addresses the questions: How do family-controlled firms pay their executives? And how do their pay practices compare with those of other firms? The dimensions of pay practice investigated are the total pay level and the relative weight of stock-based incentives. The executive pay practices of family-controlled firms are compared to those of widely held firms.

Typically, families maintain control of a publicly traded firm by holding a block of stock large enough to ensure that they can appoint the majority of the board of directors.[1] I define a family-controlled firm as one with an individual or with one or more families who own 20% or more of the voting shares of the firm. To identify a closely held firm, Morck, Strangeland, and Yeung (1998) and Park, Nelson, and Torabzadeh (2000), among others, use 20% voting control, which is the legal threshold of de facto control of a publicly traded firm in Canada. I include a firm controlled by an individual in family-controlled firms, since this individual is likely to be interested in maintaining the firm's control in his or her family. Smith and Amoako-Adu (1999) provide additional reasons for including in the definition of family firms cases where one person owns a controlling voting interest.

Using Canadian data for 1993-1997, I find that family-controlled firms place less weight on stock-based incentives than do widely held firms. I also document the empirical evidence that the reported overall executive pay is lower for family-controlled firms than for widely held firms. In the remainder of the paper, I discuss research questions, describe study methods and data, and then report the results followed by conclusions.

---

[1] See Chua, Chrisman, and Sharma (1999) for a thorough review of the dimensions of family involvement in a business organization.

Research Questions

Compensation mix

It is generally agreed that when the ownership of a publicly traded firm is dispersed among a large number of shareholders, there is little monitoring of management, providing temptation for managerial self-dealing (Smith 1776; Berle and Means 1932; Alchian and Demsetz 1972; Jensen and Meckling 1976).[2] Since the costs of monitoring are borne privately, while the benefits of monitoring are widely shared, individual shareholders in a widely held firm are unlikely to recover the cost of personally monitoring management performance. In equilibrium there would be no monitoring on the part of shareholders.

There are at least two ways to deal with the problem of managerial self-dealing: (1) the provision of incentives which align the interests of executives with those of shareholders (Jensen and Muphy 1990), and (2) the presence of large shareholders who are able to recoup their investment in monitoring (Alchian and Demsetz 1972). The use of stock-based incentives results in higher compensation costs for the firm than does fixed pay with direct monitoring, since a typical manager's investment portfolio consists largely of the human capital invested in the firm. The inability to diversify that investment means that the manager requires a greater return on the human capital invested in the firm to compensate for the non-diversifiable risk that must be assumed (Milgrom and Roberts, 1992). As a result, an incentive-based approach is more costly than one based on flat pay under direct monitoring.

In the context of family control, the cost of increased monitoring to controlling families may be more than offset by the increased valuation of their equity investment through their monitoring activity. Thus, the monitoring argument implies that since the controlling family is better off substituting direct monitoring for more expensive stock-based incentives, the use of stock-based incentives will be less important in a family-controlled firm than in a widely held firm (Hypothesis H1). Park, Nelson, and Torabzadeh (2000) suggest controlling shareholder monitoring to explain their finding that, in general, closely held firms use fewer stock-based incentives than do widely held firms.

As well, the primary objective of families who control publicly traded firms may be to maximize the family wealth rather than shareholder

---

[2] Managerial self-dealing can take the form of shirking, excessive perk consumption, overinvestment, squandering of corporate funds, overcompensation, cronyism, expropriation and theft.

wealth (Morck, Strangeland, and Yeung 1998). If controlling families reward executives for personal loyalties in growing the family wealth rather than for their contribution to the firm performance, the use of stock-based incentives will be less important in a family-controlled firm than in a widely held firm.

Size of total compensation

In addition to predicting the weight of stock-based incentives, standard agency literature predicts the level of executive pay for family-controlled firms. Agency theory, which holds that the use of executive stock options leads to a higher pay for executives, suggests that overall executive pay will be lower for family-controlled firms than for widely held firms, since family-controlled firms are expected to use fewer executive stock options (Hypothesis H2).

Further, if controlling families buy managerial loyalty by providing extensive executive perks which tend not to be disclosed, the true total pay their executives receive may not be less than that of executives of widely held firms, although the reported total executive pay of family-controlled firms will be less than that of widely held firms.

Research Design and Data

Data on executive compensation was collected from proxy statements solicited from firms for the period 1993 through 1997. Market data are from the TSE-Western database; accounting data are from Compustat Global Vantage; and ownership data are from both corporate proxy documents and Statistics Canada's Directory of Inter-corporate Ownership. The valuation of the stock option grants and stock appreciation rights (SAR) is based on the Black-Scholes formula for valuing a European call option as modified by Merton (1973) to account for dividend payout.[3]

[3] Data used to evaluate the present value of the stock option grant are the number of shares covered by the option grant (data source: proxy documents), stock price as measured on the day of the grant (data source: proxy documents), exercise price for the option grant (data source: proxy documents), time-to-expiration in years (data source: proxy documents), annual market yield on Canada bonds (data source: TSE-Western database), dividend yield estimated as dividend per share/closing stock price (data source: Global Vantage), and the annualized standard deviation of monthly stock returns (data source: TSE-Western Database). I follow the computation methodology described by Murphy (1985).

Table 1 Sample Distribution by Industry

SIC Codes	Industries	# of firms	%
Between 1 and 19	Agriculture and Natural Resources	21	25.0
Between 20 and 39	Manufacturing	31	36.9
Between 40 and 49	Transportation	7	8.3
Between 50 and 59	Retail and wholesale	7	8.3
Between 60 and 69	Financial services	13	15.5
Between 70 and 89	Service	5	6.0
Total		84	100.0

I use the following sample selection criteria: (i) detailed disclosure of the compensation package of at least one executive (out of five top paid executives) be available in the proxy documents; (ii) complete market and accounting data on the firm be available in the Global Vantage and the TSE-Western databases; and (iii) the firm's sales be not less than $5 million. In addition, I exclude firms with ownership structures other than family control and dispersed ownership. Finally, I exclude firms that do not appear throughout the 1993-1997 period. The final sample includes 420 firm-years (84 firms x 5 years). Table 1 shows the composition of the sample by industry. More than half of the companies in this sample are from the agriculture and natural resources, and manufacturing sectors. Table 2 shows the composition of the sample by ownership structure. Of 420 firm-years, family control is observed in 163 firm-years (38.8%), while ownership is widely dispersed in the remaining 257 firm-years (61.2%).

Table 3 shows executive pay and firm characteristics. Average salary, bonus, stock option and total pay in 1993 constant dollars are $265,040, $141,270, $366,000, and $772,300, respectively. Total pay was averaged over all executives with complete compensation data for a given fiscal year. Total pay includes salary, bonuses, and the value of stock options as disclosed in the proxies.[4] Mean weights of salary, bonus, and stock option in total pay are 50.00%, 17.39%, and 31.62%,

Table 2 Distribution of Ownership Structure of the Sample Firms

Ownership Type	# of firm-years	%
Family Control	163	38.8
Widely Held	257	61.2
Total	420	100.0

Table 3 Summary Statistics of the Variables

	Mean	Median	Minimum	Maximum	St. Dev.
Salary ($)	265,040	224,740	70,434	1,436,700	148,940
Bonus ($)	141,270	71,763	0	2,975,800	267,320
Stock Option ($)	366,000	131,860	0	7,030,500	698,600
Total Pay ($)	772,300	514,060	102,840	7,387,800	817,460
Weight of Salary (%)	50.99	50.08	3.07	100.00	25.06
Weight of Bonus (%)	17.39	13.44	0.00	95.87	16.43
Weight of Stock Option (%)	31.62	27.41	0.00	96.21	27.68
Market-to-Book Ratio of Assets	1.09	0.95	0.07	5.32	0.79
Sales ($million)	1,952	544	8	33,191	3,791
Annual Stock Return (%)	29.7	14.3	-85.0	1859.5	108.4

Note: Weight is averaged over all executives with complete compensation information for a given company-year. All dollar figures are adjusted to 1993 dollars. The mean weights of salary, bonus, and stock option are the average weights of salary, bonus, and stock options in total pay, respectively.

respectively.[5] Market-to-book ratio of assets ranges from 0.07 to 5.32; sales range from $8 million to $33 billion; and annual stock return ranges from -85% to 1859.5%.

To explore executive pay practices of family-controlled firms, I develop models of executive pay using the standard panel data regression analysis. In particular, I report regression results where error terms, e_{it}, are assumed to exhibit heteroskedasticity and time auto-regression following the specification by Kmenta (1986, 616-25).[6] To study the effect

[4] Other annual compensation, value of restricted shares, and long-term incentive payout disclosed in the proxies are excluded since they represent very small amounts. Other compensation disclosed in the proxies is excluded since it often includes many extraneous items such as severance pay and awards of contracts. Furthermore, when other annual compensation, value of restricted shares, and long-term incentive payout, and all other compensation are added to the total pay, I obtain essentially the same regression results.

[5] These figures (50%, 17.39%, and 31.62%) are unweighted averages, e.g., the average weights of salary, bonus, and stock options in total pay, respectively. They are not weighted averages, e.g., the ratios of the average salary, bonus, and stock options to the average total compensation of $772,300.

[6] In the model estimation, I allow for heteroskedasticity, which refers to the variability of variances across observations, since larger values of WTSO as well as LTOTCOMP are likely to have larger variances. I also allow for the autocorrelation of error terms, since the estimation errors of the same-firm observations are likely be correlated over time.

of family control on the relative use of stock-based incentives, I use the following model (1):

$$WTSO_{it} = b_0 + b_1 FAMCON_{it} + b_2 LSALES_{it} + b_3 MBRATIO_{it} + e_{it}$$

where $WTSO_{it}$ is the ratio of the value of stock option to total pay. $FAMCON_{it}$ is an indicator variable, which assumes the value of one if firm "i" is under family control and zero otherwise. I control for firm size using log of sales ($LSALES_{it}$). Gaver and Gaver (1995) find a positive relationship between the weight of stock-based incentives and firm size. I control for the firm's investment opportunities, as did Gaver and Gaver, using Chung and Pruitt's (1994) approximation of Tobin's q ($MBRATIO_{it}$).[7] Among others, Clinch (1991), Smith and Watts (1992), and Gaver and Gaver find a positive relationship between the weight of stock-based incentives and firm's investment opportunities.

To study the effect of family control on total pay, I use the following model (2):

$$LTOTCOMP_{it} = \beta_0 + \beta_1 FAMCON_{it} + \beta_2 LSALES_{it} + \beta_3 MBRATIO_{it} + \beta_4 RET_{it} + \varepsilon_{it}$$

where $LTOTCOMP_{it}$ is the log of average total compensation for firm "i" in year "t" expressed in 1993 constant dollars. RET_{it} is the annual stock return for firm "i" in year "t". Among others, Bryan and Hwang (1997), Yermack (1998), and Baker (1999) find that sales and stock return are positively correlated with total compensation. Sales is used as a proxy for firm size, while stock return for the fiscal year is used as a proxy for shareholder wealth change. I control for the firm's investment opportunities ($MBRATIO_{it}$) as in Gaver and Gaver (1995).

Results

The estimated coefficients of the models (1) and (2) used to investigate the effect of family control on executive pay practices are reported in Table 4. In both models, the widely held firms are the default class of governance structure. In both models, the sign and magnitude of the coefficient estimates for control variables (LSALES, MBRATIO, and RET) are essentially consistent with those reported in the literature.

---

[7] Chung and Priutt's approximation of Tobin's q is obtained as the ratio of the sum of market value of common stock, book value of long-term debt, book value of short-term debt, and book value of preferred stock to book value of total assets.

Compensation mix

In the model (1) column of Table 4, I relate the weight of stock options in total compensation (WTSO) to the family control (FAMCON) variable. The coefficient estimate of FAMCON is negative and significant after controlling for the impacts of size and growth opportunities. This result indicates that the proportion of stock-based incentives to total pay is smaller for family-controlled firms than for widely held firms. This is consistent with the large shareholder monitoring literature which suggests that direct monitoring in family-controlled firms would make the weight of stock-based incentives less important (H1). However, this finding is also consistent with the explanation based on the self-dealing of controlling families: if executives in family-controlled firms are paid for personal loyalties in helping controlling families to loot the firm's wealth rather than for their contribution to firm's performance, the use of stock-based incentives will be reduced.

Size of total compensation

In the model (2) column of Table 4, the natural log of total compensation (LTOTCOMP) is the dependent variable. Total compensation is higher in large firms (LSALES) and in high growth firms (MBRATIO). The estimated coefficient of annual stock returns (RET) is positive but not significant. The regression results of model (2) are consistent with family-controlled firms paying executives less than widely held firms. The coefficient estimate of FAMCON, which measures the difference in the intercept term between family-controlled firms and widely held firms, is negative and highly significant. This indicates that total compensation for family-controlled firms is significantly lower even after having controlled for size, growth, and corporate performance. This is consistent with the standard monitoring argument; since family-controlled firms use stock options as incentives less than do widely held firms and cash payment is less costly to a firm than are stock-based incentives, family-controlled firms pay their executives less than widely held firms (H2). However, this result may be an artifact of the imperfect measure of total pay used in the study, which does not explicitly include perks. If family-controlled firms use perks extensively, the measure of total pay used here, which does not explicitly include perks, may distort the difference in pay between family-controlled firms and widely held firms.

Table 4
Estimated Coefficients of Panel Data Regression Models of the Weight of Stock-based Executive Incentives and the Total Pay on Family Control

Variables	model (1)	model (2)
Intercept	17.53*** (3.83)	10.97*** (113.10)
FAMCON	-15.04*** (-8.59)	-0.08** (-2.16)
LSALES	1.05* (1.82)	0.29*** (23.97)
MBRATIO	11.79*** (8.70)	0.39*** (13.27)
RET		0.02 (0.68)
Loglikelihood	-1075.06	295.66

Note: N = 420 observations consisting of 84 firms (cross-sectional units) observed continuously for the 1993-1997 period. The error terms satisfy heteroskedasticity, cross-sectional independence, and timewise autoregression.

*** denotes p-value < 0.01
** denotes p-value < 0.05
* denotes p-value < 0.10

Conclusions

This study documents that family-controlled firms place less weight on stock-based incentives than do widely held firms. This finding can be explained using a standard monitoring argument: controlling families supply direct monitoring of management, reducing the need for stock-based incentives. On the other hand, a controlling family's incentive to reward its executives for personal loyalties in growing the family wealth may account for the finding that family-controlled firms use fewer market-based incentives than widely held firms.

This study also documents that family-controlled firms pay their executives less than widely held firms. This finding can be explained using portfolio theory argument. Standard portfolio theory implies that one dollar of stock option is less valuable to an executive whose portfolio is typically heavily concentrated on his or her own firm, than is one certain dollar. Additional compensation that accompanies stock-based incentives makes them more expensive to the firm than cash payments. Since controlling families make less use of stock-based incentives, the total executive pay of family-controlled firms will be less than that of widely held firms. On the other hand, this finding may be an artifact caused by the imperfect measure of total pay used in the study. If family-controlled firms make extensive use of perks for executive remuneration, the total pay for executives in family-controlled firms may appear to be less than the total benefits executives actually receive, since the measure of total pay used in this study does not include perks.

Disentangling the alternate explanations, refinement of empirical measure, and investigating the relationship between executive pay practices of family-controlled firms and the entrenchment of inept descendants are all interesting future avenues of research.

Acknowledgments

I am grateful to Jess Chua and Jim Chrisman, editors, Toni Nelson, Khalil Torabzadeh, Mark Huson, and Linda Janz, and two anonymous referees for their suggestions and helpful comments.

References

Alchian, A, and Demsetz, H. (1972). Production, information costs, economic organization. *American Economic Review* 62: 777-95.

Baker, T.A. (1999). Options reporting and the political costs of CEO pay. *Journal of Accounting, Auditing and Finance* 14: 125-46.

Berle, A, and Means, G. (1932). *The Modern Corporation and Private Property.* New York: Macmillan.

Bryan, S., and Hwang, L. (1997). CEO compensation in a regulatory environment: An analysis of the electric utility industry. *Journal of Accounting, Auditing and Finance* 12: 223-51.

Chua, J., Chrisman, J., and Sharma, P. (1999). Defining the family firm by behavior. *Entrepreneurship: Theory and Practice* 23(4): 19-39.

Chung, K.H., and Pruitt, S.W. (1994) A simple approximation of Tobin's q. *Financial Management* 23: 70-74.

Clinch, G. (1991). Employee compensation and firms' research and development activity. *Journal of Accounting Research* 29: 59-78.

Daniels, R.J., and Halpern, P. (1996). Too close for comfort: The role of the closely held public corporation in the Canadian economy and the implications for public policy. *Canadian Business Law Journal* 26: 11-62.

Eckbo, B., and Verma, S. (1994). Managerial shareownership, voting power, and cash dividend policy. *Journal of Corporate Finance* 1: 33-62.

Francis, D. (1986) *Controlling Interest: Who Owns Canada.* Toronto: Macmillan.

Gaver, J.J., and Gaver, K.M. (1995). Compensation policy and the investment opportunity set. *Financial Management* 24: 19-32.

Jensen, M.C., and Meckling, W.H. (1976). Theory of the firm: Managerial behavior, agency costs and ownership structure. *Journal of Financial Economics* 4: 305-60.

Jensen, M.C., and Murphy, K.J. (1990). Performance pay and top management incentives. *Journal of Political Economy* 98: 225-64.

Kmenta, J. (1986). *Elements of Econometrics, 2nd Edition.* New York: Macmillan.

Merton, R. (1973). Theory of rational option pricing. *Bell Journal of Economics and Management Science* 4: 141-83.

Milgrom, P., and Roberts, J. (1992). *Economics, Organization, and Management,* 206-47, 432-47. New Jersey, Prentice Hall.

Morck, R.K. (1996). On the economics of concentrated ownership. *Canadian Business Law Journal* 26: 63-85.

Morck, R.K., Strangeland, D.A., and Yeung, B. (1998). Inherited wealth, corporate control, and economic growth: The Canadian disease? Working Paper 6814, *National Bureau of Economic Research.*

Murphy, K.J. (1985). Corporate performance and managerial remuneration: An empirical analysis. *Journal of Accounting and Economics* 7: 11-42.

Newman, P. (1998). *Titans: How the New Canadian Establishment Seized Power.* Toronto: Penguin Books.

Park, Y.W., Nelson, T., and Huson, M. (2001). Executive pay and the disclosure environment: Canadian evidence. *Journal of Financial Research* 24: 347-65.

Park, Y.W., Nelson, T., and Torabzadeh, K. (2000). Controlling shareholder and executive incentive structure: Canadian evidence. *Canadian Journal of Administrative Sciences* 17: 245-53.

Smith, A. (1776). *An Inquiry into the Nature and Causes of Wealth of Nations.* 1976 Canon edition, New York: Modern Library.

Smith, C.W., and Watts, R.L. (1992). The investment opportunity set and corporate financing, dividend and compensation policies. *Journal of Financial Economics* 32: 263-92.

Smith, B.F. and Amoako-Adu, B. (1999). Management succession and financial performance of family controlled firms. *Journal of Corporate Finance* 5: 341-68.

Yermack, D. (1998). Companies' modest claims about the value of CEO stock option awards. *Review of Quantitative Finance and Accounting* 10: 207-26.

Chapter 16

Exploring the Role of Trust in Family Firm Governance

Lloyd Steier

Abstract

Trust plays a significant role in the governance of many organizations. For family firms in particular, trust plays an important role in organizational governance and potentially represents an important source of strategic advantage. This paper presents an overview of relevant literature on trust and organizational governance. It then explores how this literature is relevant to furthering our understanding of the management and study of family enterprise.

Trust and Organizational Governance

The creation of an effective system of governance is a central task of all organizations (Bradach and Eccles 1989; Coase 1937; Powell 1990; Sahlman 1990; Steier 1998; Stinchcombe 1965; Williamson 1975, 1985). Broadly defined, governance is the framework within which exchanges occur; it is manifest in various forms, including market, hierarchy, relational contracting, as well as "hybrid" or "plural" combinations of these (Bradach and Eccles 1989; Steier 1998). From a strategy perspective, a key managerial challenge is creating and sustaining governance structures that fit the realities of changing external and internal environments. Organizational forms that prevail "are those that deal most efficiently with the cost of economic transactions" (Grabher 1993, 3). A significant body of literature has begun to emphasize the role of trust in organizational governance (Bradach and Eccles 1989; Kramer and Tyler 1996; Powell 1987, 1990; Rousseau, Sitkin, Burt and Camerer 1998).

In other words, trust plays an important role in a firm's survival and success. Used effectively, trust represents a major source of competitive advantage for a firm. As a governance control mechanism it provides some clear advantages. "Trust is an important lubricant of a social system. It is extremely efficient; it saves people a lot of trouble to have a fair degree of reliance on other people's word" (Arrow 1974, 23). Mayer,

Davis, and Schoorman (1995, 712) provide a useful working definition of trust: "the willingness of a party to be vulnerable to the actions of another party based on the expectation that the other will perform a particular action important to the trustor, irrespective of the ability to monitor or control that other party." Researchers and practitioners alike recognize that trust potentially lowers transaction costs while contributing to more effective managerial coordination and collaboration within a firm. Conversely, distrust can lead to problems of coordination and control and consequently undermine an enterprise. There can also be a mismatching of trust where the trust of one party is not reciprocated.

Given its important role within organizations, trust should be of interest to all those concerned with a firm's performance. The actions of those responsible for the strategic direction of the organization greatly influence the level of trust within organizations. Within this context, Wicks, Berman, and Jones (1999) reinforce the notion that organizations should strive for a level of optimal trust. These authors suggest that organizations sometimes "overinvest" or "underinvest" in trust and that neither state is desirable. Too much trust can lead to problems. For example, cases of embezzlement often involve scenarios where a party has been trusted too much. Conversely, too little trust is also an undesirable state if it has to be replaced by elaborate safeguard mechanisms that greatly increase transaction costs.

Striking a balance wherein trust is at an optimal level represents a perennial management challenge. From a research and theory development perspective, further understanding the dynamics of trust within organizations represents a "central issue in the analysis of trust" (Tyler and Kramer 1996, 6).

Scholarly work in the fields of law, sociology, economics, organizational theory and strategy has developed a number of different, albeit complementary, concepts related to trust and its role in business transactions and organizational governance. The following section briefly highlights some of this work.

In a classic article written from a legal perspective, Macauley (1963, 55) makes the central observation that governance of business transactions has an important dimension that goes beyond formal agreements and contracts. He makes two central observations regarding the conduct of business: first, he observes that people "often fail to plan exchange relationships completely"; and second, they "seldom use legal sanctions to adjust these relationships and settle disputes." In other words, they tend to avoid the transaction costs associated with legal specificity and also employ non-legalistic means of imposing sanctions when problems arise. While legal contracts with boilerplate provisions remain an important aspect of the conduct of business, "not all exchanges are neatly rationalized" (Macauley1963, 58). Macauley notes that businesspeople often prefer to rely on mechanisms such as "a man's word," a "handshake," or "common honesty and decency." Lorenz (1999, 301) reinforces this view of economic exchange, suggesting that "in the real world," incomplete contracts are pervasive and social ties such as "trust, friendliness or loyalty" play an important role in market exchange. In essence, trust plays an important role in business transactions, often superseding legalistic approaches to deal creation, deal structuring, and deal termination. Sociologists further substantiate this view of economic exchange. For example, Granovetter (1985, 481) notes that most accounts of economic action provide "undersocialized" or "atomized-actor" explanations; he makes a strong case that business activity is "embedded" in structures of social relations. Similarly, Grabher (1993, Preface) observes that "various forms of networks involve intricate and durable relations which are based on trust and reciprocal patterns of communication and exchange."

The fields of economics and organizational theory further recognize trust as playing a key role in organizational governance. Building on the work of Coase (1937), Williamson (1975) articulates two fundamental forms of economic organization: market and hierarchy. Williamson, (1991) also describes a third generic form of economic organization as a "hybrid," presenting it as an "intermediate" form of governance.

Bradach and Eccles (1989, 97) observe that a "myriad of organizational forms exist along with market and hierarchy." In reiterating that transactions are rarely governed by the single forms of market and hierarchy, they emphasize that these mechanisms "serve as the building blocks for complex social structures so common in organizational life" (1989, 98). Hybrid arrangements have recently become "highly significant features of the contemporary organizational landscape" (Powell 1987, 68). Larson (1992, 76) defines them thus: "hybrids combine aspects of market transactions and characteristics of hierarchies and fall between the two alternatives on a continuum." Bradach and Eccles (1989) cite a growing body of literature that suggests relational contracting based on trust is an important governance mechanism. They further suggest that the major models of governance – market, hierarchy, and relational contracting based on trust – are not mutually exclusive and are often "combined with each other in assorted ways in the empirical world" (1989, 116). They describe this combination as a "plural form" of governance. Similarly, Ring and Van De Ven (1992, 483) include trust as a significant mode of governance, along with market and hierarchy, which is particularly useful in cases of "repeated transactions of highly idiosyncratic assets under conditions of uncertainty, and small numbers bargaining." These authors view a combination of the risk associated with the venture and the nature of the trust between the parties as key determinants of whether market, hierarchical or relational contracts based on trust are used. In other words, each form of governance has its own basis of cooperation. Markets largely rely on self-interest, hierarchies rely on coercion, and relational contracting relies on trust.

Every form of governance must address the problem of a party's tendency to behave opportunistically. Relational contracting is particularly prone to opportunistic behavior and relies on a strong element of trust as the basis for cooperation. When trust is embedded in the relationship, opportunistic behavior is unlikely to occur because each partner will forego short-term gains in favor of the interests of the partnership. Bradcach and Eccles (1989) suggest that, for trust to be meaningful, there must be a risk of "opportunistic" behavior.

Relational views of competitive advantage (Dyer and Singh 1998) are also receiving increased attention within the field of strategic management. This perspective focuses on "dyad/network routines and processes as an important unit of analysis for understanding competitive advantage" (1998, 661). Relational exchanges are built over time and reflect an ongoing process. They are characterized by "purposeful cooperation" and "complex webs" of operational and social interdependence (Gundlach and Murphy 1993, 36). In essence, substantial barriers to competition can be created through "structural" and "social" bonding. From an economic development perspective, some of the more dynamic regions throughout the world are characterized by institutions that facilitate trust (Lorenz 1992). Indeed for many countries, trust, residing in institutions such as the family, provides important sources of competitive advantage (Fukuyama 1995).

Trust reduces the transaction costs associated with exchange because it reduces monitoring costs while providing safeguards against opportunistic behavior. Trust is embedded in social relationships (Granovetter 1985). It is usually generated over time (Granovetter 1985,1992; Lorenz 1999). Repeated experiences of successful interaction usually result in an incremental accumulation of trust. In terms of organizations and economic transactions, "embedded social relationships" (Fiet 1995, 197), "personal trust" (Williamson 1993); and "habitualization" (Nooteboom, Berger, and Nooderhaven 1997) often substitute for more formalized governance arrangements.

In summary, business relationships provide participating parties with opportunities to take advantage of one another. They contain an element of relational risk. Close relationships, or ties that bind, build trust and represent a primary source of sustained competitive advantage because they reduce relational risk. Trust is an important mechanism governing economic transactions and organization (Mayer, Davis and Schoorman 1995; Bradach and Eccles 1989; Ring and Van De Ven 1992; Rousseau, Sitkin, Burt, and Camerer 1998). For most firms, trust represents an important, albeit little understood, attribute of organization. Family firms, in particular, represent a type of organization wherein the dynamics of trust potentially plays a most important role in firm survival and success.

Family firms as unique organizational forms and the potential dynamics of trust

Family businesses represent unique organizational forms with dimensions that go beyond the bottom line; consequently, business transactions are rarely purely economic. In the phrase "family businesses," the word "family" implies something other than "business" is manifest within the organization. For these firms, trust often represents a fundamental basis for cooperation and potentially provides a key source of competitive advantage. In matters of organizational governance, too much – or too little – trust can be problematic. While family enterprises are challenged to develop governance mechanisms that permit the building and sustaining of optimal trust, relatively little is known about the dynamics of trust within these unique organizational forms. By definition, the notion of "family" implies bonds that, if used effectively, can provide a source of strategic advantage for a firm.

The term "family business" has many meanings (Hoy and Verser 1994; Litz 1995). While the term is lacking in "definitional clarity" (Litz 1995), there is a clear consensus that family businesses evolve through various stages. This paper assumes a very broad definition of family business to include that group of persons affiliated by common ancestry or marriage who have a financial stake in a business enterprise. While all business transactions involve relationships, family businesses involve additional ties that bind beyond the business relationship. Within this context, a family business might involve siblings, parents and children, cousins, marriage partners (in some cases, ex-partners) and in-laws.

Clearly, family firms represent dynamic entities that grow and change over time. Their evolution is well documented. Change has been the subject of considerable attention by organizational scholars (Baum 1996). A central observation of this research is that change is a naturally occurring process. As already noted, a critical task of those concerned with firm performance is to develop governance structures appropriate for the firm as it evolves through various stages of development. For example, stages models, combined with a focus on the transitions between each stage, provide a useful schema for examining a variety of phenomena. These models are often used within the context of family enterprise, and a number of models exist (Neubauer and Lank 1998; Ward 1991). Ward provides a schema that is well-established in the business literature and is a particularly useful illustration for our purposes. He depicts family businesses as experiencing three stages, each characterized by its dominant form of ownership: Founding Owners, Sibling Partnership, and Confederation of Cousins.

Another key feature of the stages literature on organizations is the emphasis on transition periods. Transitions are particularly important because organizations often experience predictable crises. Organizations that do not address the challenges associated with the transition can be limited in their growth or even fail. Greiner's (1972) often-cited article "Evolution and Revolution as Firms Grow" describes four growth stages of firm development and the crisis that accompanies each stage:

Punctuated equilibrium models of change (Gersick 1991) also reinforce the notion that long periods of incremental change are interrupted by discontinuous radical change. For example, within a family firm a change in the ownership structure might signal radical change.

Gersick, Davis, Hampton and Lansberg (1997; also see Gersick, Lansberg, Desjardins, and Dunn 1999) expand on previously discussed existing models with their introduction of a "three dimensional" development model of family business. They represent these dimensions as occurring along three axes: (1) a family axis that includes stages of: the young business family, entering the business, working together, and passing the baton; (2) a business axis that includes stages of: start-up, expansion/formalization, and maturity; and (3) an ownership axis that includes stages of: controlling owner, sibling partnership, and cousin consortium.

Much of the literature on trust suggests that it grows over time. However, the stages literature relative to family business implies processes implicit in family firm evolution that may diminish trust. Simple growth

Table 1
Greiner's Evolution and Revolution in Firms

Stage	Crisis
1) Entrepreneurial	Leadership
2) Collectivity	Need for delegation and control
3) Formalization	Too much red tape
4) Elaboration	Bureaucracy has reached its limits; need for revitalization and social control.

of the family unit implies a dilution of the ties that bind. Other experiences that a family may commonly encounter also impact the level of trust (e.g., intergenerational conflict, sibling rivalry, divorce, multiple marriages, cousin disputes). Accordingly, alternative governance structures must be created that supplement or replace trust as a governance mechanism. Conversely, there may also be instances where family firms rely too little on trust as a governance mechanism and thereby miss an important source of strategic advantage. In other words, as family firms evolve so should the role of trust as a governance mechanism.

Discussion and Conclusions

The aim of this article has been to explore the dynamics of trust and its relationship to organizational governance within the family firm. Conceptually, it is based on the notion that trust and the nurturing of trust-building processes should be of fundamental concern to family firms. While family firms can be characterized as having specific mechanisms that facilitate trust, there are other forces that lead to the dilution or dissolution of trust. At times these forces can be so powerful that they create an environment of distrust.

Interestingly, the evolution of trust as a governance mechanism within family firms may exhibit an evolutionary pattern distinctly different from that of non family firms. A significant body of literature on business relationships, joint ventures and strategic alliances suggests that relationships start with formal contracts, market controls, or hierarchical governance systems and, as parties get to know one another better, are gradually replaced by relational contracts which rely on trust as the basis of cooperation. Family firms appear to rely a great deal on trust as a governance mechanism during their early stages of development and often become challenged to introduce additional governance mechanisms as they grow and evolve.

Stages models of family firm evolution suggest that there are forces at play that may diminish trust. Familial ties that bind may gradually become weaker and unleash forces which diminish, rather than build, trust. For example, later generations "may not trust other family members in the same way they trusted their parents" (Drozdow and Carroll 1997, 75). As family firms evolve, trust – which may have been a major governance feature and significant source of competitive advantage – may be out of alignment with the change. Within family firms it is not unusual for "strategic imperatives" and "family life cycle events" to be "out of phase" (Drozdow and Caroll 1997). The dilution of trust may be a naturally occurring process. Those concerned with firm performance must recognize a misalignment when it occurs and then follow an appropriate combination of the following strategies: (1) reinvest in trust and trust-building activities, and/or (2) introduce supplementary, more plural forms of governance.

This article has suggested that trust plays an important role in the governance of family firms. There has been relatively little research about the dynamics of trust within these unique organizations. Future research must focus on a number of issues; however, it is first necessary to gather data to further determine the validity of the concepts presented in this paper. Some focal questions for further research include: Is a trust a potential source of strategic advantage for family firms over non-family firms? How do family firms use trust as a governance mechanism? Are trust contracts among members of family firms more robust than in non-family firms? If trust represents a major source of strategic advantage for family firms what is the impact on these organizations when major betrayals of trust occur? Does the optimal role of trust evolve as firms experience different stages of development? What are the conditions that foster optimal trust? As family firms evolve, do they naturally experience situations wherein distrust occurs? What are the conditions that foster distrust?

In conclusion, it is hoped that this paper has provided both insights and ideas for action to those involved in the management of family firms as well as those concerned with furthering our understanding of these firms through further research.

References

Arrow, K. (1974). *The Limits of Organization.* New York: Norton.

Baum, J.A.C. (1996). Organizational ecology. In *Handbook of Organizational Studies*, ed. S. R. Clegg, C. Hardy, and W. Nord, 77-114. London: Sage.

Bradach, J. L., and Eccles, R. G. (1989). Price, authority, and trust: From ideal types to plural forms. *Annual Review of Sociology* 15: 97-118.

Coase, R. (1937). The nature of the firm. *Economica* 4: 386-405.

Drozdow, N., and Carroll, V.P. (1997). Tools for strategy development in family firms. *Sloan Management Review* 39(1): 75-88.

Dyer, J. H., and Singh, H. (1998). The relational view: Cooperative strategy and sources of interorganizational competitive advantage. *Academy of Management Review* 23: 660-79.

Fiet, J.O. (1995). Reliance upon informants in the venture capital industry. *Journal of Business Venturing* 10: 195-223.

Fukuyama, F. (1995). *Trust, the Social Virtues and the Creation of Prosperity*. New York: Free Press.

Gersick, C. (1991). Revolutionary change theories: A multilevel exploration of the puncturated equilibrium paradigm. *Academy of Management Review* 16: 10-36.

Gersick, K. E., Davis, J.A., Hampton, M.M., and Lansberg, I. (1997). *Generation to Generation: Life cycles of the Family Business*. Boston: Harvard Business School Press.

Gersick, K. E., Lansberg, I. Desjardins, M. and Dunn, B. (1999). Stages and transitions: Managing change in the family business. *Family Business Review* 12: 287-97.

Grabher G. (Ed.) (1993). *The Embedded Firm: On the Socioeconomics of Industrial Networks*. London: Routledge & Kegan Paul.

Granovetter, M. S. (1985). Economic action and social structure: The problem of embeddedness. *American Journal of Sociology* 91: 481-510.

Granovetter, M. (1992). Problems of explanation in economic sociology. In *Networks and Organizations: Structure, Form, and Action*, ed. N. Nohria and R.G. Eccles, 25-56. Boston: Harvard Business School Press.

Greiner, L. (1972). Evolution and revolution as firms grow. *Harvard Business Review* 50(4): 37-46.

Gundlach, G.T., and Murphy, P. E. (1993). Ethical and legal foundations of relational marketing exchanges. *Journal of Marketing* 57: 35-36.

Hoy, F., and Verser, T. G. (1994). Emerging business, emerging field: Entrepreneurship *Entrepreneurship Theory and Practice* 19(1): 9-23.

Kramer, R.M., and Tyler, T. R. (1996). *Trust in Organizations: Frontiers of Theory and Research.* Thousand Oaks, California: Sage.

Larson, A. (1992). Network dyads in entrepreneurial settings: A study of the governance of exchange relationships. *Administrative Science Quarterly* 37: 76-104.

Litz, R. A. (1995). The family business: Toward definitional clarity. *Academy of Management Proceedings*: Vancouver.

Lorenz, E. H. (1992). Trust, community, and cooperation: Toward a theory of industrial districts. In *Pathways to Industrialization and Regional Development*, ed. M.Storper and A.J. Scott, 195-204. Routledge: New York.

Lorenz, E.H. (1999). Trust, contract and economic cooperation. *Cambridge Journal of Economics* 23: 301-315.

Macauley, S. (1963). Non-contractual relations in business: A preliminary study. *American Sociological Review* 28: 55-67.

Mayer, R. C., Davis, J. H., and Schoorman, F. D. (1995). An integrative model of organizational trust. *Academy of Management Review* 20: 709-734.

Neubauer, F. and Lank, A. (1998). *The Family Business. Its Governance for Sustainability*. London: MacMillan Press Ltd.

Nooteboom, B., Berger, H., and Noorderhaven, N.G. (1997). Effects of trust and governance on relational risk. *Academy of Management Journal* 40: 308-338.

Powell, W.W. (1987). Hybrid organizational arrangements. *California Management Review* 30: 67-87.

Powell, W.W. (1990). Neither market nor hierarchy: Network forms of organization. In *Research in Organizational Behavior*, ed. B.M. Staw and L.L. Cummings, 12, 295-396. Greenwich, CT:JAI Press.

Ring, P.S. and Van De Ven, A.H. (1992). Structuring cooperative relationships between organizations. *Strategic Management Journal* 13:483-98.

Rousseau, D.M., Sitkin, S.B., Burt, R.S., and Camerer, C.C. (1998). Not so different after all: A cross-discipline view of trust. *Academy of Management Review* 23: 393-405.

Sahlman, W.A. (1990). The structure and governance of venture-capital organizations. *Journal of Financial Economics* 27: 473-521.

Steier, L. (1998). Confounding market and hierarchy in venture capital governance: The Canadian immigrant investor program. *Journal of Management Studies* 35: 511-35.

Stinchcombe, A.L. (1965). Social structure and organizations. In *Handbook of organizations*, ed. J.G. March, 142-93. New York: Rand McNally.

Tyler, T.R.and Kramer, R.M. (1996). "Whither trust?" In *Trust in organizations: Frontiers of theory and research,* ed. T.R. Tyler and R.M. Kramer, 1-15. Thousand Oaks, California: Sage.

Ward, J.L. (1991). *Creating Effective Boards for Private Enterprises: Meeting the Challenges of Continuity and Competition.* San Francisco: Jossey-Bass.

Wicks, A.C., Berman, S.L., and Jones, T.M. (1999). The structure of optimal trust: Moral and strategic implications. *Academy of Mangement Review* 24: 99-116.

Williamson, O.E. (1975). *Markets and Hierarchies: Analysis and Anti-trust Implications.* New York: Free Press.

Williamson, O.E. (1985). *The Economic Institutions of Capitalism.* New York: Free Press.

Williamson, O.E. (1991). Comparative economic organization: The analysis of discrete structural alternatives. *Administrative Science Quarterly* 36, 269-96.

Williamson, O. E. (1993). Calculativeness, trust, and economic organization. *Journal of Law & Economics* 36: 453-86.

Notes on Contributors

Michael W. Chiasson is an Assistant Professor of Management Information Systems in the Faculty of Management of the University of Calgary. His research interests encompass a broad range of agency-structure issues in the initiation, design, development and dissemination of information systems — particularly e-commerce and medical information systems. His current research projects include software systems design diffusion and infusion, knowledge management, interpersonal conflict, project management, social theory, e-commerce law, health promotion and medicine, history of technology, philosophy, and qualitative research methods — particularly action research.

James J. Chrisman is a Professor of Entrepreneurship and Innovation in the Faculty of Management at the University of Calgary. He holds an Endowed Professorship in Family Business Entrepreneurship and is Director of the Centre for Family Business Management and Entrepreneurship. His primary research interests are the strategic management of family business, the impact of entrepreneurial assistance programs, and corporate entrepreneurship.

Jess H. Chua is a Professor of Finance in the Faculty of Management at the University of Calgary. He holds an Endowed Professorship in Family Business Management. His primary research interests are the strategic management of family business and financial management.

Vance Gough is an Instructor at Mount Royal College. His MBA in Enterprise Development at the University of Calgary in 1995 led him to create the Prairie Mill Bread Company, which expanded to three locations before being sold in 2001. From 1996-2000 he also worked as an instructor and Senior Development Officer of the Project Centre in the Faculty of Management of the University of Calgary, where he provided initial front line consultation to hundreds of small Calgary businesses and taught Opportunity Identification courses and seminars for other organizations, including TransCanada Pipelines and the Alberta Provincial Government.

J. Adam D. Holbrook is an Adjunct Professor and Associate Director at the Centre for Policy Research on Science and Technology (CPROST), at Simon Fraser University in Vancouver, B.C. Prior to his appointment at CPROST he was involved in S&T policy activities in a number of federal government departments in Ottawa. At CPROST his research activities have centred on the analysis of science, technology and innovative activities in both the public sector and the private sector. He is the leader of InnoCom, a network of researchers in innovation studies in western Canada, and the western regional director for the national network of researchers on innovation systems and industrial clusters, the Innovation Systems Research Network.

Lindsay P. Hughes has degrees in electrical engineering and communication. He recently completed his MA in communication at Simon Fraser University, during which the research on which his chapter is based was completed. He is now Project Manager at Sage Research in Natick, Massachusetts, USA.

Peter Josty is currently Chair of the Steering Team for the Centre for Innovation Studies in Calgary. He has 30 years of industrial experience in such areas as research and development, marketing, technical management, new application development, new business development and strategic planning. He has a PhD in chemistry from the University of London, and an MBA from IMI/University of Geneva, with specialization in technological innovation.

Robert Kleysen is a PhD student at the I.H. Asper School of Business at the University of Manitoba. His research interests include entrepreneurship, organizational learning, innovation, and business ethics. His forthcoming articles will appear in *Busness Ethics Quarterly*, *Journal of Intellectual Capital*, and *Family Business Review*.

Cooper H. Langford is Director of the Science, Technology, and Society Program and Professor of Chemistry at the University of Calgary. He is former Vice-President (Research) at the University of Calgary and former chair of the MettNet board. He is active in the Innovation Systems Research Network.

Reginald A. Litz is an Associate Professor at the I. H. Asper School of Business at the University of Manitoba. His research focuses on small firm strategy and family business management and has been published in several leading outlets of entrepreneurship research including *Journal of Business Venturing*, *Entrepreneurship Theory and Practice*, and *Journal of Small Business Management*. After work he enjoys piano improvisation and composition, which helped enormously in conceptualizing his article.

W. Ed McMullan is a Professor of Entrepreneurship and Innovation in the Faculty of Management at the University of Calgary. He has written approximately 60 articles published in a variety of journals such as *Journal of Business Venturing, Entrepreneurship Theory and Practice, Technovation*, and *Journal of Small Business Management*. He is also co-author of the book, *Developing New Ventures: The Entrepreneurial Option* (1990), which introduces a midrange theory of entrepreneurship. His research interests are in the area of entrepreneurship education. He is founder and major stockholder of Polymer Sciences Inc., a Calgary-based company that develops and manufacturers adhesives and coatings for the construction industry.

Ron Mitchell is an Associate Professor of Entrepreneurship and holder of the Francis G. Winspear Chair in Public Policy and Business at the University of Victoria. He also holds a joint appoinment in the Department of Strategy and Public Policy at the Guanghua School of Management at Peking University, Beijing, PRC. His research interests focus on increasing individual economic independence in society—both domestically and internationally—through the study of entrepreneurship and the development of transaction cognition theory.

Eric Morse is an Assistant Professor of Entrepreneurship in the Faculty of Business at the University of Victoria. His principle areas of expertise include entrepreneurial strategy, corporate entrepreneurship and entrepreneurial cognition. Dr. Morse is a co-designer of the UVic Entrepreneurship Program and also teaches the capstone strategy class for the MBA Program. He is an instructor for the school's International Executive MBA Program and President/Director of the International Centre for Venture Expertise. He also serves on the board of directors for several local start-ups. His past work experience includes time with Andersen Consulting and the Los Alamos National Laboratory – in its technology transfer office.

James K. O'Grady is an Associate Professor and Director of the Industrial Design Program in the Faculty of Environmental Design at the University of Calgary. He is also completing a PhD. in Entrepreneurship and Innovation in the Faculty of Management at the University of Calgary. Mr. O'Grady has extensive international experience as a design consultant and educator, having practised, taught and studied design and design management in Canada, the United States and Great Britain.

Yun W. Park is an Associate Professor of finance at the University of Lethbridge, Alberta.. His research centers around corporate governance issues highlighting resolutions of conflict of interest in firms. More recently, his work on the executive compensation in Canada appeared in *Canadian Journal of Administrative Sciences* and *Journal of Financial Research.*

Janet Peace was the founder, President, and CEO of MettNet. She continues to work on knowledge transfer issues in her consulting practice, Janet Peace Associates. She has an ongoing connection with Dalhousie University and is Partnerships and Business Development Director of iDLab.

Peter W.B. Phillips is a Professor and NSERC/SSHRC Chair in Managing Knowledge-based Agri-food Development at the University of Saskatchewan. His research concentrates on intellectual property management for agricultural biotechnology, trade and marketing issues related to GM foods and development strategies. Dr. Phillips is a member of the Canadian Biotechnology Advisory Committee, Senior Research Associate at the Estey Centre for Law and Economics in International Trade and Co-Director of the Canadian Agri-food Trade Research Consortia.

Pramodita Sharma is an Associate Professor at Wilfrid Laurier University. She is a recipient of various national and international research-related awards, including the prestigious 1998 NFIB Best Dissertation Award from the Academy of Management, and Best Paper Awards from Family Firm Institute (FFI), Canadian Council of Small Business and Entrepreneurship (CCSBE), and Administrative Sciences Association of Canada (ASAC). Her research interests are in family business.

Robert A. (Sandy) Slator is President and Chief Executive Officer of PTI Group Inc., North America's leading integrated provider of lodging facilities and food services to remote locations. PTI serves, primarily, the natural resource sector in Canada and the Gulf of Mexico. Prior to joining PTI Group Inc., he served as President and CEO of Vencap Equities Alberta Ltd., at that time Canada's largest venture capital company. He also served as President of the Canadian Venture Capital Association from 1991 until 1994.

Richard Smith is an Associate Professor of Communication in the Faculty of Applied Science in Simon Fraser University, Burnaby, British Columbia. He is also Director of the Centre for Policy Research on Science and Technology (CPROST) at Simon Fraser University at Harbour Centre, Vancouver, British Columbia. He is currently involved in research that looks at: (1) critical policy studies of new applications for information technology equipment and services in support of teamwork and distance education for universities as well as the primary and secondary school environment; and (2) the role of social capital in the management of innovation and collaborative R&D in start-ups and small high technology enterprises.

Lloyd Steier is an Associate Professor in the Department of Strategic Management and Organization at the University of Alberta School of Business. He is also affiliated with the U of A's Centre for Entrepreneurship and Family Enterprise as an endowed professor and academic director. Dr. Steier has researched and published in the areas of strategic management, new venture creation, venture capital governance, management education, and family enterprise.

Anne Tyrie was a senior MettNet Associate. She holds a PhD in geophysics and has been active in the information technology industry. She has carried forward some of the MettNet activity in her present firm, People Interactive, in Calgary.

Karl H. Vesper is a Professor of Business Administration, Mechanical Engineering and Marine Studies at the University of Washington, Seattle. He has held endowed professorships in entrepreneurship as a visitor at Baylor University, Babson College and the University of Calgary and served as a Fullbright Distinguished Lecturer at Trinity College, Dublin. At Babson he helped create the Center for Entrepreneurial Studies in 1980 and also started the annual Babson Entrepreneurship Research Conference and its publication *Frontiers of Entrepreneurship Research*, which is now in its 20th year. His latest book is *Opportunity Discovery* (Korean, with Byung Ju Cho), ChungAh publishing, 1999.

Myron Weber is an Associate Professor of Human Resources and Organization Dynamics in the Faculty of Management at the University of Calgary and has also served as Executive Director of the Institute of Health Economics in Alberta. His research and teaching interests encompass the areas of the management and economics of health care and human resources management and include health promotion, the economics of HIV/AIDS, risk-taking and decision-making, and the cost-effectiveness of wellness programs.

Index

C

K

L

M